Ohio State University Mathematical Research Institute Publications 9

Editors: Gregory R. Baker, Walter D. Neumann, Karl Rubin

Ohio State University
Mathematical Research Institute Publications

1 Topology '90, *B. Apanasov, W. D. Neumann, A. W. Reid, L. Siebenmann* (Eds.)
2 The Arithmetic of Function Fields, *D. Goss, D. R. Hayes, M. I. Rosen* (Eds.)
3 Geometric Group Theory, *R. Charney, M. Davis, M. Shapiro* (Eds.)
4 Groups, Difference Sets, and the Monster, *K. T. Arasu, J. F. Dillon, K. Harada, S. Sehgal, R. Solomon* (Eds.)
5 Convergence in Ergodic Theory and Probability, *V. Bergelson, P. March, J. Rosenblatt* (Eds.)
6 Representation Theory of Finite Groups, *R. Solomon* (Ed.)
7 The Monster and Lie Algebras, *J. Ferrar, K. Harada* (Eds.)
8 Groups and Computation III, *W. M. Kantor, Á. Seress* (Eds.)

Complex Analysis and Geometry

Proceedings of a Conference
at the Ohio State University
June 3–6, 1999

Editor

Jeffery D. McNeal

Walter de Gruyter · Berlin · New York 2001

Editor
Jeffery D. McNeal
Department of Mathematics, The Ohio State University, Columbus, OH 43210, USA

Series Editors
Gregory R. Baker
Department of Mathematics, The Ohio State University, Columbus, OH 43210-1174, USA
Karl Rubin
Department of Mathematics, Stanford University, Stanford, CA 94305-2125, USA
Walter D. Neumann
Department of Mathematics, Columbia University, New York, NY 10027, USA

Mathematics Subject Classification 2000: 32–06; 32A07, 32A10, 32A40, 32G08, 32G15, 32H02, 32H35, 32T25, 32T27, 32W05

Keywords: Pseudoconvex domains, Proper mappings, $\bar{\partial}$-Neumann problem, The $\bar{\partial}$-problem, hypoellipticity, Levi-flat hypersurfaces

Printed on acid-free paper which falls within the guidelines of the ANSI to ensure permanence and durability.

Library of Congress – Cataloging-in-Publication Data

Complex analysis and geometry : proceedings of a conference at the Ohio State University, June 3–6, 1999 / editor, Jeffery D. McNeal.
p. cm. – (Ohio State University Mathematical Research Institute publications ; 9)
ISBN 311016809X (alk. paper)
1. Mathematical analysis – Congresses. 2. Functions of complex variables – Congresses. 3. Geometry – Congresses. I. McNeal, Jeffery D. II. Series.
QA299.6 .C6577 2001
515–dc21
2001047181

Die Deutsche Bibliothek – Cataloging-in-Publication Data

Complex analysis and geometry : proceedings of a conference at the Ohio State University, June 3–6, 1999 / ed. Jeffery D. McNeal. – Berlin ; New York : de Gruyter, 2001
(Ohio State University Mathematical Research Institute publications ; 9)
ISBN 3-11-016809-X

Printed in Germany.
Cover design: Thomas Bonnie, Hamburg.
Typeset using the authors' $\mathrm{T_EX}$ files: I. Zimmermann, Freiburg.
Printing and binding: Hubert & Co. GmbH & Co. KG, Göttingen.

Preface

A conference on Complex Analysis and Geometry was held at the Ohio State University from June 3 to June 6, 1999. This volume contains ten articles written by some of the principal speakers at this conference.

The conference was an exciting event and showcased some of the new ideas which continue to fuel the strong interaction between analysis and geometry in several complex variables. The articles are mostly oriented toward researchers in the field, but the authors were invited to include more expository and illustrative material in their papers than they might include in a normal journal article. It is hoped, therefore, that the volume will also serve as an introduction to students to some of the active areas in complex analysis.

Besides the contributors to this volume, I would like to thank the many mathematicians who attended the conference and made it such a stimulating event. I also gratefully acknowledge the support of the National Science Foundation and the Mathematical Research Institute at Ohio State, which helped make the conference possible. Finally, I thank Manfred Karbe and Annette Kölbl of Walter de Gruyter & Co. for their professional and patient help in bringing this book to completion.

J. D. McNeal

Table of contents

Points in general position in real-analytic submanifolds in $\mathbb{C}^N$ and applications

M. S. Baouendi, Linda Preiss Rothschild and Dmitri Zaitsev

1. Introduction

Two pairs (M, p) and (M', p') of germs of real (locally closed) submanifolds $M, M' \subset \mathbb{C}^N$ at distinguished points $p \in M$ and $p' \in M'$ are said to be *biholomorphically equivalent* (or just *equivalent* for short) if there is a biholomorphic map H between open neighborhoods of p and p' in $\mathbb{C}^N$ sending p to p' and mapping a neighborhood of p in M onto a neighborhood of p' in M'. We write $(M, p) \sim (M', p')$ for equivalent pairs and $H: (\mathbb{C}^N, p) \to (\mathbb{C}^N, p')$ for a map between open neighborhoods of p and p' in $\mathbb{C}^N$ sending p to p'.

It is easy to construct germs of smooth real curves in $\mathbb{C}$ that are not equivalent. In contrast, any two germs of *real-analytic* curves in $\mathbb{C}$ at arbitrary distinguished points are always equivalent, since any real-analytic diffeomorphism between them extends to a biholomorphism between some open neighborhoods in $\mathbb{C}$.

The simplest example of non-equivalent real-analytic submanifolds of the same dimension is given by $(\mathbb{C}, 0)$ and $(\mathbb{R}^2, 0)$ both linearly embedded in $\mathbb{C}^2$ in the standard way. More generally, it is easy to see that two germs at 0 of real linear subspaces of $\mathbb{C}^N$ are equivalent if and only if they can be transformed into each other by a complex linear automorphism of $\mathbb{C}^N$.

In this paper we give a local description of a real-analytic submanifold $M \subset \mathbb{C}^N$ at a "general" point (see Theorem 2.5 below). This description is based on various notions of nondegeneracy and is of interest in its own right. An important application is that at a "general" point $p \in M$, the germ (M, p) is equivalent to another germ (M', p') if and only if (M, p) and (M', p') are "formally" equivalent (see Theorem 6.1 below.) This result is the main theorem in **[BRZ 2000]**. (It should be noted that there exist pairs (M, p) and (M', p') which are "formally" equivalent, but not biholomorphically equivalent; see §4.1.) We also address here the case of real-algebraic submanifolds and their algebraic equivalences, which was not studied in **[BRZ 2000]**. (See Theorem 9.1 below.)

We mention briefly that the study of biholomorphic equivalence of real submanifolds in $\mathbb{C}^N$ goes back to Poincaré **[P 1907]** and E. Cartan **[Ca 1932a]**, **[Ca 1932b]**, and **[Ca 1937]**. In their celebrated work Chern and Moser **[CM 1974]** solved the equivalence problem for germs of Levi nondegenerate real-analytic hypersurfaces in

Complex Analysis and Geometry
Ohio State Univ. Math. Res. Inst. Publ. 9

$\mathbb{C}^N$ and showed in particular that in this context the notions of formal and biholomorphic equivalence coincide. We will mention more recent work related to Theorem 6.1 later in this article.

2. Structure decomposition results for points in general position

If M is a connected real-analytic submanifold of $\mathbb{C}^N$, we say that a property holds for $p \in M$ in *general position* if it holds for all p outside a proper real-analytic subvariety of M. A number associated to M is said to be a *biholomorphic invariant* if it is preserved by biholomorphic equivalences of germs of M at any point in M.

2.1. Generic and CR submanifolds. A smooth real submanifold $M \subset \mathbb{C}^N$ is called *generic* (or *generating*, in some translations) if $T_pM + JT_pM = T_p\mathbb{C}^N$ for all $p \in M$, where $J\colon T\mathbb{C}^N \to T\mathbb{C}^N$ denotes the standard complex structure of $\mathbb{C}^N$, and T_pM denotes the (real) tangent space of M at p. More generally, if the space $T_qM + JT_qM$ has constant dimension for q near p, M is said to be CR at p (or p is a CR point of M). If M is CR at every point, it is said to be a CR submanifold. (For CR manifolds, the reader is referred e.g. to the books **[J 1990]**, **[Bo 1991]**, **[Ch 1991]**, **[BER 1999a]**.) If M is real-analytic, the set of all non CR points of M is a nowhere dense proper real-analytic subvariety of M.

Examples 2.1. In $\mathbb{C}$ all nontrivial smooth submanifolds are generic. More generally, the graph of any (smooth) map between open sets in $\mathbb{C}^n \times \mathbb{R}^{N-n}$ $(0 \le n \le N)$ and $i\mathbb{R}^{N-n}$ is generic. For $N \ge 2$, complex submanifolds of $\mathbb{C}^N$ of positive codimension and their real submanifolds are never generic. The submanifold

$$M := \{w = |z|^2\} \subset \mathbb{C}^2,$$

where (z, w) are taken as coordinates in $\mathbb{C}^2$, is generic and CR everywhere except at the origin in $\mathbb{C}^2$.

The role of generic points is illustrated by the following property.

Proposition 2.1. *If $M \subset \mathbb{C}^N$ is a connected real-analytic submanifold, there exists an integer $0 \le r_1 \le N$ such that for $p \in M$ in general position*

$$(M, p) \sim (M_1 \times \{0\}, 0), \quad M_1 \times \{0\} \subset \mathbb{C}^{N-r_1} \times \mathbb{C}^{r_1}, \tag{2.1}$$

where $M_1 \subset \mathbb{C}^{N-r_1}$ is a generic real-analytic submanifold through 0. The number r_1 with this property is unique and is a biholomorphic invariant.

Remark 2.2. The points p for which the conclusion of Proposition 2.1 holds are in fact the CR points of M.

The number r_1 is called the *excess codimension* of M (cf. **[BRZ 2000]**, §2). It is equal to the maximal codimension of a complex submanifold of $\mathbb{C}^N$ containing an open subset of M. For $p \in M$ in general position, M is CR at p and there exists a complex submanifold of $\mathbb{C}^N$ of codimension r_1 that contains a neighborhood of p in M. This complex submanifold is unique in the sense of germs and is called the *intrinsic complexification* of M at p.

2.2. Finite and minimum degeneracy. Finite nondegeneracy is a higher order generalization of Levi nondegeneracy. For a smooth CR submanifold $M \subset \mathbb{C}^N$ we denote by T^cM the real subbundle given by $T^c_pM = \{X \in T_pM : JX \in T_pM\}$. We consider the (0, 1) *vector fields* on M, i.e., the sections of the subbundle

$$T^{0,1}M := \{X + iJX : X \in T^cM\} \subset T^cM \otimes \mathbb{C}.$$

Then M is *Levi nondegenerate* at p if for any (0, 1) vector field L with $L(p) \neq 0$, there exists a (0, 1) vector field L_1 such that

$$[L_1, \overline{L}](p) \notin T^c_pM \otimes \mathbb{C}.$$

This condition is equivalent to the nondegeneracy of the Levi form defined as the (unique) hermitian form

$$\mathcal{L}_p : T^{0,1}_pM \times T^{0,1}_pM \to (T_pM/T^c_pM) \otimes \mathbb{C} \tag{2.2}$$

satisfying

$$\mathcal{L}_p(L_1(p), L(p)) = \frac{1}{2i}\pi[L_1, \overline{L}](p) \tag{2.3}$$

for all (0, 1) vector fields L, L_1, where $\pi : TM \otimes \mathbb{C} \to (TM/T^cM) \otimes \mathbb{C}$ is the canonical projection.

The more general concept of finite nondegeneracy can be defined in a similar way as follows.

Definition 2.1. A smooth CR submanifold $M \subset \mathbb{C}^N$ is called *finitely nondegenerate* at p if there exists $l \geq 0$ such that for any (0, 1) vector field L on M with $L(p) \neq 0$, there are (0, 1) vector fields on M, $L_1, \dots, L_k$, $0 \leq k \leq l$, such that

$$[L_1, \dots, [L_k, \overline{L}] \dots](p) \notin T^c_pM \otimes \mathbb{C}. \tag{2.4}$$

If l is minimal with this property, M is called *l-nondegenerate* at p.

Remark 2.3. It follows that M is 1-nondegenerate if and only if it is Levi-nondegenerate. Also M is 0-nondegenerate if and only if it is totally real, i.e., $T^c_pM = \{0\}$. To check Definition 2.1 it suffices to assume that $L, L_1, \dots, L_k$ in (2.4) are all taken from a fixed local basis of (0, 1) vector fields on M near p.

The condition of finite nondegeneracy was given in **[BHR 1996]** for hypersurfaces and can be found in **[BER 1999a]** for CR submanifolds of higher codimension. The formulation given in Definition 2.1 is equivalent to that in the reference above (see § 3 or Proposition 1.24 in **[E 2000]**). In **[BRZ 2000]** this notion was extended to arbitrary real-analytic submanifolds. As in the case of the Levi form, higher order tensors can be used to give an alternative definition of finite nondegeneracy (see §3).

Example 2.4. Let $M \subset \mathbb{C}^2$ be a real-analytic hypersurface through 0 given by $\operatorname{Im} w = \phi(z, \bar{z}, \operatorname{Re} w)$ where ϕ is a real-valued real-analytic function defined near the origin in $\mathbb{C} \times \mathbb{R}$ satisfying $\phi(z, 0, 0) \equiv 0$. Then M is finitely nondegenerate at 0 if and only if at least one of the partial derivatives $\phi_{z,\bar{z}^k}(0)$ $(1 \leq k < \infty)$ does not vanish. If k is the smallest such integer, then M is k-nondegenerate at 0.

The role of finite nondegeneracy is illustrated by the following property.

Proposition 2.2. *If $M \subset \mathbb{C}^N$ is a real-analytic submanifold, then there exists an integer $0 \leq r_2 \leq N$ such that for $p \in M$ in general position*

$$(M, p) \sim (M_2 \times \mathbb{C}^{r_2}, 0), \quad M_2 \times \{0\} \subset \mathbb{C}^{N-r_2} \times \mathbb{C}^{r_2}, \tag{2.5}$$

where $M_2 \subset \mathbb{C}^{N-r_2}$ is a real-analytic CR submanifold finitely nondegenerate at 0. The number r_2 with this property is unique and is a biholomorphic invariant.

The number r_2 is called the *degeneracy* of M. It is equal to the maximal dimension of the leaves of a holomorphic foliation in a neighborhood of a point $p \in M$ such that a neighborhood of p in M is saturated (i.e., is a union of leaves) (see **[F 1977]**). The points $p \in M$ for which the conclusion of Proposition 2.2 holds are said to be of *minimum degeneracy* in M (see **[BRZ 2000]**, §2).

Example 2.5. The hypersurface $M \subset \mathbb{C}^3$ given by $\operatorname{Im} w = |z_1 z_2|^2$ in the coordinates (z_1, z_2, w) is of minimum degeneracy 1 outside the plane $H := \{(0, 0)\} \times \mathbb{R}$. For $p \notin H$, (M, p) is equivalent to $(M_1 \times \mathbb{C}, 0)$ with $M_1 := \{(z, w) \in \mathbb{C}^2 : \operatorname{Im} w = |z|^2\}$.

2.3. Finite type and CR orbits. A smooth CR submanifold $M \subset \mathbb{C}^N$ is said to be of *finite type* at p (in the sense of Kohn **[K 1972]** and Bloom–Graham **[BG 1977]**) if all $(0, 1)$ vector fields on M and their conjugates, together with all their higher order commutators span the space $T_pM \otimes \mathbb{C}$. For general CR submanifolds, the condition of finite nondegeneracy and that of finite type are independent, i.e., one condition does not necessarily imply the other. (See Examples 2.6 and 2.7.) However, for hypersurfaces finite nondegeneracy at a point implies finite type at that point.

Example 2.6. The hypersurface $M \subset \mathbb{C}^2$ given in Example 2.4 is of finite type at 0 if and only if at least one of the partial derivatives $\phi_{z^k\bar{z}^l}(0)$, $1 \leq k, l < \infty$, does not vanish. Hence, for instance, the hypersurface given by

$$M = \{(z, w) \in \mathbb{C} \times \mathbb{C} : \operatorname{Im} w = |z|^4\}$$

is of finite type at 0, but is not finitely nondegenerate at 0.

Example 2.7. For any Hermitian form $\langle \, , \rangle : \mathbb{C} \times \mathbb{C} \to \mathbb{C}^2$ the quadric

$$M: \; = \{(z, w) \in \mathbb{C} \times \mathbb{C}^2 : \operatorname{Im} w = \langle z, z \rangle\}$$

is not of finite type at any point. If, in addition, the vector-valued Hermitian form $\langle z, z \rangle$ does not vanish identically, then M is finitely nondegenerate at every point.

Proposition 2.3. *If $M \subset \mathbb{C}^N$ is a connected real-analytic submanifold, then there exists an integer $0 \le r_3 \le N$ such that for $p \in M$ in general position*

$$(M, p) \sim (M_3, 0), \quad M_3 \subset \mathbb{C}^{N-r_3} \times \mathbb{R}^{r_3}, \tag{2.6}$$

with $(0, u) \in M_3$ for $u \in \mathbb{R}^{r_3}$ sufficiently small, and for such fixed u, $M_3 \cap (\mathbb{C}^{N-r_3} \times \{u\})$ is a CR submanifold of finite type at $(0, u)$. The number r_3 with this property is unique and is a biholomorphic invariant.

The number r_3 is called the *orbit codimension.* The *CR orbit* of a point $p \in M$ is the germ at p of a (real-analytic) submanifold of M through p of smallest possible dimension to which all the $(0, 1)$ vector fields on M are tangent. The number r_3 is equal to the minimal codimension of any CR orbit. It is also equal to the maximal number r such that there exists a holomorphic submersion of a neighborhood of a point $p \in M$ in $\mathbb{C}^N$ onto $\mathbb{C}^r$ sending M into the real part $\mathbb{R}^r$. (See **[BER 1996]** for the discussion in the algebraic case.) The points p for which the conclusion of Proposition 2.2 holds are said to be of *minimum orbit codimension* (see **[BRZ 2000]**, §2). The germ of the submanifold $M_3 \cap (\mathbb{C}^{r_3} \times \{u\})$ at $(0, u)$ is the CR orbit of M_3 at this point.

In contrast to the situation of Proposition 2.2, in general the submanifold M_3 cannot be "flattened", i.e., written as $M_3' \times \mathbb{R}^{r_3}$, with $M_3' \subset \mathbb{C}^{N-r_3}$ a real-analytic submanifold. One obstruction to such a "flattening" is the possible nonequivalence of the CR orbits $M_3 \cap (\mathbb{C}^{N-r_3} \times \{u\})$ for different $u \in \mathbb{R}^{r_3}$. In order to give an example of such a submanifold that cannot be "flattened" at any point, we introduce some preliminaries.

Consider for any real number u the quadric $M_u \subset \mathbb{C}^6$ of codimension 2 defined in the coordinates $(z, w) \in \mathbb{C}^4 \times \mathbb{C}^2$ by $\operatorname{Im} w = h_u(z, z)$, where h_u is the $\mathbb{C}^2$-valued Hermitian form

$$h_u(z, \chi) := \left(z_1\overline{\chi}_1 + z_3\overline{\chi}_3 + z_4\overline{\chi}_4, \;\; z_2\overline{\chi}_2 + z_3\overline{\chi}_3 + u z_4\overline{\chi}_4\right). \tag{2.7}$$

The following lemma will be used to construct our example.

Lemma 2.4. *For any $u, u' \in \mathbb{R}$ with $u, u' > 1$ the Hermitian forms h_u and $h_{u'}$ are equivalent if and only if $u = u'$. That is, if there exist a 2×2 invertible real matrix B*

and a 3×3 *complex invertible matrix* A *such that*

$$Bh_{u'}(z, \chi) = h_u(Az, A\chi) \quad \textit{for all } z, \chi \in \mathbb{C}^4, \tag{2.8}$$

then $u = u'$.

Proof. Let $\{e_1, \ldots, e_4\}$ be the standard basis of $\mathbb{C}^4$. We claim first that if $\{v_1, \ldots, v_4\}$ is a basis of $\mathbb{C}^4$ for which $h_u(v_j, v_k) = 0$, $j \neq k$, then there is a permutation σ of $\{1, 2, 3, 4\}$ and complex numbers $\lambda_j \neq 0$ such that $v_j = \lambda_j e_{\sigma(j)}$, $j = 1, \ldots, 4$. To prove the claim, let ϕ_u be the positive definite (scalar) Hermitian form given by $\phi_u(z, \chi) = h_u^1(z, \chi) + h_u^2(z, \chi)$, where h_u^1, h_u^2 are the components of h_u. Then the 4×4 diagonal matrix D with diagonal entries $\big(1, 0, 1/2, 1/(1+u)\big)$ satisfies

$$\phi_u(Dz, \chi) = h_u^1(z, \chi) \quad \text{for all } z, \chi \in \mathbb{C}^4. \tag{2.9}$$

Hence $\phi_u(Dv_j, v_k) = 0$, $j \neq k$, and since ϕ_u is positive definite, it follows that $Dv_j = l_j v_j$, for some $l_j \in \mathbb{C}$, $j = 1, \ldots, 4$. In particular, since D is diagonal and its eigenvalues are distinct, any eigenvector of D is a nonzero multiple of one of the e_j. Since the v_j are all eigenvectors of D, the claim is proved.

Assume that (2.8) holds. Since $h_{u'}(e_j, e_k) = 0$, $j \neq k$, we may apply the claim to the vectors $v_j = Ae_j$ and conclude that $Ae_j = \lambda_j e_{\sigma(j)}$, $j = 1, \ldots, 4$. From this and (2.8) applied to $z = \chi = e_j$, we have $Bh_{u'}(e_j, e_j) = |\lambda_j|^2 h_u(e_{\sigma(j)}, e_{\sigma(j)})$, $j = 1, \ldots, 4$. Hence the real 2×2 invertible matrix B maps each vector in the set $S_{u'} := \{(1,0), (1,1), (1,u'), (0,1)\}$ into a positive multiple of a vector in the set $S_u := \{(1,0), (1,1), (1,u), (0,1)\}$. Since each of the vectors $(1,1)$ and $(1,u')$ is a linear combination with positive coefficients of the vectors $(1,0)$ and $(0,1)$, the same is true of their images under the linear map B. It follows that each of the vectors $(1,0)$ and $(0,1)$ is mapped into a positive multiple of one of the same two vectors. Similar reasoning shows that up to positive scalar multiples B sends the vectors in the set $S_{u'}$ into those of S_u either by preserving the order of these vectors or reversing it. In either case, a simple calculation shows that necessarily $u = u'$. □

As a consequence of Lemma 2.4, it follows that for any $u, u' \in \mathbb{R}$, $u, u' > 1$ and $p \in M_u$, $p' \in M_{u'}$ the germs (M_u, p) and $(M_{u'}, p')$ are equivalent if and only if $u = u'$. Indeed since $(M_u, p) \sim (M_u, 0)$ for any $p \in M_u$ and any $u \in \mathbb{R}$, it suffices to assume $p = p' = 0$. If $(M_u, 0)$ and $(M_{u'}, 0)$ are (biholomorphically) equivalent, their Levi forms h_u and $h_{u'}$ are linearly equivalent, i.e., they must satisfy (2.8). Hence $u = u'$ by the lemma. Moreover, for any $u \in \mathbb{R}$ and any $p \in M_u$, the reader can easily check that M_u is of finite type at p. We may now give an example of a manifold which cannot be "flattened" at any point, as announced above.

Example 2.8. Let $h_u(z, z)$ be defined by (2.7). Consider the generic submanifold $M \subset \mathbb{C}^7$ of codimension 3 given in the coordinates $(z, w, u) \in \mathbb{C}^4 \times \mathbb{C}^2 \times \mathbb{C}$ by

$$\operatorname{Im} u = 0, \quad \operatorname{Im} w = h_u(z, z).$$

Then the CR orbits of M are $M \cap (\mathbb{C}^6 \times \{u\})$, $u \in \mathbb{R}$. By the observation preceding this example, the CR orbits of M are pairwise nonequivalent for $u > 1$. Hence, for any $q = (p, u) \in M$, with $p \in M_u$, $u > 1$, (M, q) is not equivalent to any product $(M' \times \mathbb{R}, 0)$ with $M' \subset \mathbb{C}^6$ a CR submanifold of finite type. Indeed, the CR orbit of $(0, u')$ in $M' \times \mathbb{R}$ is $M' \times \{u'\}$ and hence such orbits are equivalent to each other for different values of u'.

2.4. A structure result for points in general position. Putting Propositions 2.1, 2.2 and 2.3 together we obtain:

Theorem 2.5. *Let $M \subset \mathbb{C}^N$ be a connected real-analytic submanifold. Then there exist integers $0 \le r_1, r_2, r_3 \le N$, such that for $p \in M$ in general position*

$$(M, p) \sim (\widetilde{M} \times \mathbb{C}^{r_2} \times \{0\}, 0), \qquad \widetilde{M} \times \mathbb{C}^{r_2} \times \{0\} \subset \left(\mathbb{C}^r \times \mathbb{R}^{r_3}\right) \times \mathbb{C}^{r_2} \times \mathbb{C}^{r_1}, \quad (2.10)$$

where $r := N - r_1 - r_2 - r_3$, $\widetilde{M} \subset \mathbb{C}^r \times \mathbb{R}^{r_3}$ is a finitely nondegenerate generic submanifold of $\mathbb{C}^{r+r_3}$ through 0 such that for $u \in \mathbb{R}^{r_3}$ near 0, the point $(0, u)$ is in $\widetilde{M}$ and $\widetilde{M} \cap (\mathbb{C}^r \times \{u\})$ is a CR submanifold of finite type at $(0, u)$.

2.5. The hypersurface case. In the case $M \subset \mathbb{C}^N$ is a hypersurface, Theorem 2.5 can be reformulated in a simpler form. First, M is generic in $\mathbb{C}^N$, so that $r_1 = 0$. Also, as mentioned in §2.3, for $N \ge 2$, any hypersurface which is finitely nondegenerate at 0 is necessarily of finite type at 0, so that $r_3 = 0$. Hence we have

Corollary 2.6. *Let $M \subset \mathbb{C}^N$ be a connected real-analytic hypersurface. Then exactly one of the following alternatives holds.*

(a) *There exists an integer $0 \le r_2 \le N - 2$ such that for all $p \in M$ in general position, $(M, p) \sim (\widetilde{M} \times \mathbb{C}^{r_2}, 0)$, where $\widetilde{M} \subset \mathbb{C}^{N-r_2}$ is a finitely nondegenerate hypersurface (and hence of finite type) through 0.*

(b) *For all $p \in M$ in general position, $(M, p) \sim (\mathbb{R} \times \mathbb{C}^{N-1}, 0)$.*

3. Finite nondegeneracy and higher order tensors for CR manifolds

Let $M \subset \mathbb{C}^N$ be a smooth CR submanifold, and $p \in M$. The following construction generalizes the Levi form as given by (2.2). For $s \ge 1$, consider the linear subspace $F_p^s \subset T_p^{0,1} M$ of all possible values $L(p)$ of a $(0, 1)$ vector field L such that (2.4) fails to hold for all $k < s$ and all (0, 1) vector fields $L_1, \ldots, L_k$. We obtain a decreasing sequence of subspaces

$$T_p^{0,1}M = F_p^1 \supset \cdots \supset F_p^s \supset \cdots .$$

Then there exists a unique multilinear map

$$\mathcal{L}_p^s \colon (T_p^{0,1}M)^s \times F_p^s \to (T_pM/T_p^cM) \otimes \mathbb{C}$$

satisfying

$$\mathcal{L}_p^s(L_1(p), \ldots, L_s(p), L(p)) = \pi[L_1, \ldots, [L_s, \overline{L}] \ldots](p), \tag{3.1}$$

for all (0, 1) vector fields $L, L_1, \ldots, L_s$ with $L(p) \in F_p^s$, where π is as in (2.3). Indeed, it follows from the construction of the subspaces F_p^s that the right-hand side of (3.1) is multilinear in its arguments over the ring of smooth complex functions. The tensor $\mathcal{L}_p^s$ is complex-linear in the s first arguments and antilinear in the last one. For $s = 1$ we obtain a multiple of the Levi form; i.e., $\mathcal{L}_p^1(X_1, X_2) = 2i\mathcal{L}_p(X_1, X_2)$.

The tensors analogous to $\mathcal{L}_p^s$ were introduced by Ebenfelt **[E 1998]**. For $s = 2$, the definition of $\mathcal{L}_p^2$ is due to Webster **[W 1995]** (in the case M is a hypersurface), where $\mathcal{L}_p^2$ is called the *cubic form*. The subspaces F_p^s and the tensors $\mathcal{L}_p^s$ are also related to the submodules N_s, $1 \le s < \infty$, of the C^∞-module of all (0, 1) vector fields defined by Freeman **[F 1977]** inductively as follows. Let N_1 be the module of all (0, 1) vector fields on M. If N_{s-1} is defined, let $N_s \subset N_{s-1}$ be the C^∞-submodule consisting of all (0, 1) vector fields L such that (2.4) fails to hold for all $k < s$, all (0, 1) vector fields $L_1, \ldots, L_k$ *and all* $p \in M$. (Actually, Freeman's submodules are the conjugates of the N_s's defined here.) If all subspaces F_q^s have constant dimension for q in a neighborhood of p, N_s consists precisely of their sections. However, in general, a subspace F_p^s may be nontrivial even if $N_s = 0$. For instance, this happens for $s = 2$ if the Levi form $\mathcal{L}_p$ is nondegenerate for p in general position but degenerate at some point.

Example 3.1. The hypersurface $M \subset \mathbb{C}^2$ given by $\operatorname{Im} w = |z|^4$ is finitely degenerate on $\{0\} \times \mathbb{R}$ and finitely nondegenerate outside this line. For $p = 0$ we have $F_0^s = T_0^{0,1}M$ for all $s \ge 1$. On the other hand, the submodules N_s are all zero for $s \ge 2$ and thus "don't notice" the degeneracy.

We have the following characterization of k-nondegeneracy:

Proposition 3.1. *If $M \subset \mathbb{C}^N$ is a smooth CR submanifold and $p \in M$, then the following are equivalent for $k \ge 1$:*

(i) *M is k-nondegenerate at p,*

(ii) *k is the smallest integer for which $F_p^{k+1} = 0$,*

(iii) $F_p^k \neq \{0\}$ *and for any* $X \in F_p^k$, $X \neq 0$, *there exist* $X_1, \ldots, X_k \in T_p^{0,1}M$ *with* $\mathcal{L}_p^k(X_1, \ldots, X_k, X) \neq 0$.

We now give an equivalent definition of finite nondegeneracy in terms of a defining function of M. If $M \subset \mathbb{C}^N$ is a smooth CR submanifold of codimension d and $p \in M$, a (smooth) *defining function* for M near p is a real smooth map $\rho = (\rho^1, \ldots, \rho^d)$ of rank d defined in a neighborhood of p in $\mathbb{C}^N$ such that M is locally defined near p by $\rho(Z, \overline{Z}) = 0$. We write $\rho^j_Z = (\rho^j_{Z_1}, \ldots, \rho^j_{Z_N})$ for the complex gradient of ρ^j, $1 \leq j \leq d$. We have the following.

Proposition 3.2. *Let* $M \subset \mathbb{C}^N$ *be a smooth CR submanifold and* ρ *a smooth defining function of* M *near* $p \in M$. *Then for an integer* $k \geq 0$, *the following are equivalent.*

(i) M *is* k*-nondegenerate at* p.

(ii) *The integer* k *is minimal such that the collection of vectors* $(L_1 \ldots L_s \rho^j_Z)(p, \bar{p})$ *span* $\mathbb{C}^N$ *for* $1 \leq j \leq d$, $0 \leq s \leq k$, *and all choices of* $(0, 1)$ *vector fields* $L_1, \ldots, L_s$ *on* M.

As mentioned in Remark 2.3 above, k-nondegeneracy at p can be checked using any local basis of $(0, 1)$ vector fields on M. Also, one can directly check that condition (ii) of Proposition 3.2 does not depend on the choice of the defining function ρ and that in this condition, if suffices to take the vector fields $L_1, \ldots, L_s$ from a fixed local basis of $(0, 1)$ vector fields on M near p. To check the equivalence of (i) and (ii) in the case of a generic submanifold, the reader can take "normal" coordinates as in **[BER 1999a]** and do the calculation using the basis of vector fields given by (11.2.18) in **[BER 1999a]** and the defining function ρ given there. (A related calculation is done in the reference cited here.)

4. Different notions of equivalences for real submanifolds

4.1. Formal equivalence. In practice, biholomorphic equivalence of two germs of real submanifolds in $\mathbb{C}^N$ can be hard to check. For instance, in the work of Chern–Moser-**[CM 1974]** the so-called *formal equivalence* is established first and then the convergence of the formal map is proved, yielding biholomorphic equivalence.

Definition 4.1. A *formal equivalence* between two germs, (M, p) and (M', p'), of real-analytic submanifolds in $\mathbb{C}^N$ is an invertible formal power series map

$$H(Z) = p' + \sum_{|\alpha| \geq 1} a_\alpha (Z - p)^\alpha, \quad a_\alpha \in \mathbb{C}^N, \quad Z = (Z_1, \ldots, Z_N), \tag{4.1}$$

sending M into M' in the "formal sense", i.e., satisfying

$$\rho'\big(H(Z(x)), \overline{H(Z(x))}\big) = 0 \tag{4.2}$$

for some real-analytic parametrization $x \mapsto Z(x)$ of M near $p = Z(0)$ and some real-analytic defining function $\rho'(Z, \overline{Z})$ of M' near p'.

Equality in (4.2) is understood in the sense of formal power series in x. The formal map H given by (4.1) is invertible if the vectors $\{\partial H/\partial Z_j(p), 1 \le j \le N\}$ are linearly independent in $\mathbb{C}^N$. It is not hard to see that if (4.2) holds for some choice of parametrization $Z(x)$ of M and defining function ρ' of M', then it holds for any other choice. We shall say that the germs (M, p) and (M', p') are *formally equivalent* if there exists a formal equivalence between them. Formal equivalence of (M, p) and (M', p') translates into the existence of solutions for an infinite system of polynomial equations which the coefficients of H must satisfy. Formal equivalence is often easier to check than biholomorphic equivalence. One of the main results in **[BRZ 2000]** (see Theorem 6.1 below) states that for points in general position these two notions of equivalence coincide. Moreover, it is shown that any given formal equivalence can be "corrected" to a convergent one without changing terms of a prescribed finite order.

The assumption that the point p be in general position cannot be dropped. Indeed, in $\mathbb{C}^2$ there exist a pair of germs of formally, but not biholomorphically, equivalent 2-dimensional real-analytic (non CR) submanifolds. Moser and Webster proved in **[MW 1983]**, Proposition 6.1, that no neighborhood of 0 in the 2-dimensional (non CR) submanifold $M \subset \mathbb{C}^2$ given by

$$w = |z|^2 + \gamma\bar{z}^2 + \gamma z^3\bar{z}$$

can be biholomorphically transformed into the hyperplane $\mathbb{C} \times \mathbb{R}$ for $\gamma > 1/2$. On the other hand, they show that M is formally equivalent to a 2-dimensional real-analytic submanifold contained in $\mathbb{C} \times \mathbb{R}$ provided γ is not exceptional, i.e., if $(1/\pi)\arccos(1/2\gamma)$ is not a rational number. The authors of the present paper are not aware of any example of pairs of germs of real-analytic CR submanifolds that are formally but not biholomorphically equivalent.

4.2. CR equivalence and k-equivalence. A *CR function* on a smooth CR submanifold $M \subset \mathbb{C}^N$ is a smooth (C^∞) complex-valued function defined on M satisfying the Cauchy–Riemann equations restricted to M. More precisely, f is CR on M if $Lf \equiv 0$ for every (0, 1) vector field L on M. If M and M' are germs of smooth CR submanifolds of $\mathbb{C}^N$ at p and p' respectively, we say that (M, p) is *CR equivalent* to (M', p') if there is a CR diffeomorphism (a diffeomorphism whose components are CR functions) between open neighborhoods of p in M and of p' in M' respectively and taking p to p'. Such a diffeomorphism is called a *CR equivalence*.

It is known that if f is a CR function defined in a neighborhood of p in M, then there is a formal (holomorphic) power series $\sum_\alpha c_\alpha(Z-p)^\alpha$, $Z = (Z_1, \dots, Z_N)$, $c_\alpha \in \mathbb{C}$, whose restriction to M coincides with the Taylor series of f at p. Moreover, if M is generic, then such a formal power series is unique. (See e.g. **[BER 1999a]**, Proposition 1.7.14 for the generic case.) It follows that if h is a CR equivalence between two real-analytic germs (M, p) and (M', p') of CR submanifolds of $\mathbb{C}^N$, then the corresponding vector-valued formal power series H of the form (4.1) (obtained from the components of h by the Taylor series property of CR functions mentioned above) satisfies (4.2) and can be assumed to be invertible; hence H is a formal equivalence between (M, p) and (M', p'). On the other hand, the restriction to M of a biholomorphic equivalence between (M, p) and (M', p') is obviously a CR equivalence. Thus the notion of CR equivalence lies between that of formal and biholomorphic equivalence:

$$\text{biholomorphic equivalence} \implies \text{CR equivalence} \implies \text{formal equivalence.}$$

A weaker notion than that of formal equivalence is that of k-equivalence for an integer $k > 1$.

Definition 4.2. For an integer $k > 1$ we say that two germs, (M, p) and (M', p'), of real-analytic submanifolds of $\mathbb{C}^N$ are *k-equivalent* if there exists a biholomorphic map H between neighborhoods of p and p' in $\mathbb{C}^N$, with $H(p) = p'$, such that

$$\rho'\big(H(Z(x)), \overline{H(Z(x))}\big) = O(|x|^k)$$

for some real-analytic parametrization $x \mapsto Z(x)$ of M near $p = Z(0)$ and some real-analytic defining function $\rho'(Z, \overline{Z})$ of M' near p'. Such an H is called a *k-equivalence* between (M, p) and (M', p').

Again here the definition of k-equivalence is independent of the choice of the parametrization $Z(x)$ of M and of the choice of the defining function ρ' of M'. We also note that if H is a k-equivalence (or a formal equivalence), by taking its Taylor polynomial of order $k - 1$, we can find another k-equivalence whose components are polynomials. Hence in Definition 4.2 we could have assumed that H is a biholomorphism with polynomial components. Similarly, we could have also assumed that H is just a formal invertible mapping, rather than a biholomorphism. Then any formal equivalence may be considered as a k-equivalence for every k.

Example 4.1. If $k > 0$ is an integer, then the identity map is a $2k$-equivalence between the germs at 0 of the real hyperplane $M := \mathbb{C} \times \mathbb{R} \subset \mathbb{C}^2$ and the hypersurface M' given by $\operatorname{Im} w = |z|^{2k}$. However, it is easily checked that there is no formal equivalence between $(M, 0)$ and $(M', 0)$.

The example shows that even very different looking germs of submanifolds can be k-equivalent for some fixed k without being formally equivalent. The situation becomes rather different if we require (M, p) and (M', p') to be k-equivalent for every k. This means the existence of a sequence of biholomorphic maps H_k each sending (M, p) into (M', p') up to order k, as in Definition 4.2. In particular, as noted above, formal equivalence implies the existence of such a sequence. On the other hand, given a sequence of k-equivalences, in general, one cannot put them together to obtain a formal equivalence. Nevertheless, for points in general position, the main result in **[BRZ 2000]** states that the existence of such a sequence implies that (M, p) and (M', p') are formally and even biholomorphically equivalent. The authors of the present paper are not aware of any example of pairs of germs (M, p) and (M', p') of real analytic submanifolds in $\mathbb{C}^N$ which are k-equivalent for every $k > 1$, but not formally equivalent.

5. Structure decompositions and k-equivalences

In this section we consider the extent to which the decompositions given by Propositions 2.1, 2.2 and 2.3 (and summarized in Theorem 2.5) are invariant under different notions of equivalences.

We first consider invariance under biholomorphic equivalences. We already remarked that the numbers r_1, r_2, r_3 introduced in §2 are biholomorphic invariants. Write $r := N - r_1 - r_2 - r_3$ for brevity. Now assume that we have two germs at 0 of real-analytic submanifolds in $\mathbb{C}^N$ of the form $\widetilde{M} \times \mathbb{C}^{r_2} \times \{0\}$ and $\widetilde{M}' \times \mathbb{C}^{r_2} \times \{0\}$ with $\widetilde{M}, \widetilde{M}' \subset \mathbb{C}^r \times \mathbb{R}^{r_3}$ as in Theorem 2.5. That is, both $\widetilde{M}$ and $\widetilde{M}'$ are finitely nondegenerate generic submanifolds through 0 containing all points of the form $(0, u)$ for $u \in \mathbb{R}^{r_3}$ near 0, and such that $\widetilde{M} \cap (\mathbb{C}^r \times \{u\})$ and $\widetilde{M}' \cap (\mathbb{C}^r \times \{u\})$ are of finite type for u small. We fix local holomorphic coordinates $Z = (Z^0, Z^3, Z^2, Z^1) \in \mathbb{C}^r \times \mathbb{C}^{r_3} \times \mathbb{C}^{r_2} \times \mathbb{C}^{r_1}$ near 0 and similarly we write $H = (H^0, H^3, H^2, H^1)$ for the components of a $\mathbb{C}^N$-valued map H.

Proposition 5.1. *Let $H : (\mathbb{C}^N, 0) \to (\mathbb{C}^N, 0)$ be a biholomorphic equivalence between the germs at 0 of $\widetilde{M} \times \mathbb{C}^{r_2} \times \{0\}$ and $\widetilde{M}' \times \mathbb{C}^{r_2} \times \{0\}$. Then we have:*

(i) $H^1(Z^0, Z^3, Z^2, 0) \equiv 0$, *i.e., H sends $\mathbb{C}^r \times \mathbb{C}^{r_3} \times \mathbb{C}^{r_2} \times \{0\}$ into itself;*

(ii) $\frac{\partial H^0}{\partial Z^2}(Z^0, Z^3, Z^2, 0) \equiv 0$, $\frac{\partial H^3}{\partial Z^2}(Z^0, Z^3, Z^2, 0) \equiv 0$, *i.e., H preserves the affine subspaces given by*

$$Z^1 = 0, \quad (Z^0, Z^3) = \text{const};$$

(iii) $\frac{\partial H^3}{\partial Z^0}(Z^0, Z^3, Z^2, 0) \equiv 0$, *i.e., H also preserves the affine subspaces given by*

$$Z^1 = 0, \quad Z^3 = \text{const}.$$

In particular, the restriction of (H^0, H^3) to $\mathbb{C}^r \times \mathbb{C}^{r_3} \times \{0\} \times \{0\}$ is a biholomorphic equivalence between $(\widetilde{M}, 0)$ and $(\widetilde{M}', 0)$.

Remark 5.1. Proposition 5.1 can be reformulated "geometrically" as follows. *A biholomorphic equivalence between real-analytic submanifolds preserves their intrinsic complexifications, maximal tangent holomorphic foliations and CR orbits.*

A statement similar to Proposition 5.1 also holds for formal equivalences. However, both statements for formal and biholomorphic equivalences are in fact special cases of more general invariance properties under k-equivalences. In the following proposition, as was illustrated by Example 4.1, it is crucial to require the existence of a k-equivalence for every k.

Proposition 5.2 (**[BRZ 2000]**, Proposition 4.1). *Suppose that (M, p) and (M', p') are k-equivalent for all $k > 1$. Then the numbers r_1, r_2, r_3 in Theorem 2.5 for M coincide with the ones for M'.*

Proposition 5.3 (**[BRZ 2000]**, Lemma 4.4, Lemma 5.3). *Under the assumptions of Proposition 5.1 suppose that $\widetilde{M}'$ is l-nondegenerate and let H be a k-equivalence between $\widetilde{M} \times \mathbb{C}^{r_2} \times \{0\}$ and $\widetilde{M}' \times \mathbb{C}^{r_2} \times \{0\}$. Then we have:*

(i) $H^1(Z^0, Z^3, Z^2, 0) = O(|Z|^k)$;

(ii) $\frac{\partial H^0}{\partial Z^2}(Z^0, Z^3, Z^2, 0) = O(|Z|^{k-l-1})$, $\frac{\partial H^3}{\partial Z^2}(Z^0, Z^3, Z^2, 0) = O(|Z|^{k-l-1})$ *provided $k > l$;*

In particular, the restriction of (H^0, H^3) to $\mathbb{C}^r \times \mathbb{C}^{r_3} \times \{0\} \times \{0\}$ is a k-equivalence between $(\widetilde{M}, 0)$ and $(\widetilde{M}', 0)$.

6. Comparison of different notions of equivalences

The following, which is one of the main results of **[BRZ 2000]**, states that the four notions of equivalence discussed above actually coincide at all points in general position.

Theorem 6.1 (**[BRZ 2000]**, Corollary 14.1). *Let $M \subset \mathbb{C}^N$ be a connected real-analytic submanifold. Then for any $p \in M$ in general position and any germ (M', p') of a real-analytic submanifold in $\mathbb{C}^N$, the following conditions are equivalent:*

(i) *(M, p) and (M', p') are k-equivalent for all $k > 1$;*

(ii) *(M, p) and (M', p') are formally equivalent;*

(iii) *(M, p) and (M', p') are CR equivalent;*

(iv) *(M, p) and (M', p') are biholomorphically equivalent.*

As mentioned in §4, the implications (iv) $\Longrightarrow$ (iii) $\Longrightarrow$ (ii) $\Longrightarrow$ (i) hold trivially. It was shown in **[BER 1999b]** that if M and M' are real-analytic generic submanifolds which are finitely nondegenerate and of finite type at p and p' respectively, then any formal equivalence H between (M, p) and (M', p') is necessarily convergent. In particular, one obtains the equivalence of conditions (ii), (iii) and (iv) for M a connected real-analytic generic submanifold which is finitely nondegenerate and of finite type at some point (and hence at all points in general position).

In the case that M is a real-analytic *hypersurface*, one can use the result in **[BER 1999b]** mentioned above together with Proposition 5.3 to prove the equivalence of (ii), (iii), and (iv) of Theorem 6.1 as follows. We begin with the structure theory, Corollary 2.6, for hypersurfaces at points in general position. Since the fact that (ii) $\Longleftrightarrow$ (iii) $\Longleftrightarrow$ (iv) in case (b) of Corollary 2.6 can be easily proved, we may assume that condition (a) of that corollary holds. Hence we may assume that $(M, p) = (\widetilde{M} \times \mathbb{C}^{r_2}, 0)$ and that $(M', p') = (\widetilde{M}' \times \mathbb{C}^{r_2}, 0)$, where $\widetilde{M}$ and $\widetilde{M}'$ are finitely nondegenerate hypersurfaces and hence of finite type at 0. Let H be a formal equivalence between (M, p) and (M', p'). By Proposition 5.3 (here $r_1 = r_3 = 0$, $r = N - r_2$, and $H = (H^0, H^2)$), we conclude that the restriction $\widetilde{H}$ of H^0 to $\mathbb{C}^r \times \{0\}$ is a formal equivalence between $(\widetilde{M}, 0)$ and $(\widetilde{M}', 0)$. Since $\widetilde{M}$ and $\widetilde{M}'$ are finitely nondegenerate and of finite type at 0, it follows from the result in **[BER 1999b]** mentioned above that $\widetilde{H}$ must be already convergent. It is then easy to extend $\widetilde{H}$ to a holomorphic equivalence between $(\widetilde{M} \times \mathbb{C}^{r_2}, 0)$ and $(\widetilde{M}' \times \mathbb{C}^{r_2}, 0)$. In fact, one may choose such a biholomorphic equivalence in such a way that its Taylor series coincides with that of H up to any preassigned order.

For submanifolds of higher codimension it is not possible to reduce to the results of **[BER 1999b]**, even to prove that the notions of formal and biholomorphic equivalence coincide, since the submanifold $\widetilde{M}$ given in Theorem 2.5 need not be equivalent to a product of a CR manifold of finite type and $\mathbb{R}^m$ for some m. Indeed, if M is the finitely nondegenerate generic submanifold of codimension 3 in $\mathbb{C}^7$ given in Example 2.8, then $M = \widetilde{M}$, but $\widetilde{M}$ is not equivalent to such a product.

We now present some of the ideas involved in the proof of (ii) $\Longrightarrow$ (iv) in Theorem 6.1, as given in **[BRZ 2000]**, for a general real-analytic submanifold M. By making use of Theorem 2.5 and Proposition 5.1, we first reduce to the case where $p = p' = 0$, $M = \widetilde{M}$ and $M' = \widetilde{M}'$, with $\widetilde{M}$ and $\widetilde{M}'$ as in Theorem 2.5. (That is, we assume $r_1 = r_2 = 0$ in Theorem 2.5.)

The next step is to obtain a parametrization of all formal equivalences H between $(M, 0)$ and $(M', 0)$ by their (formal) jets along the linear subspace $C := \{0\} \times \mathbb{C}^{r_3} \subset \mathbb{C}^{N-r_3} \times \mathbb{C}^{r_3}$, which is transversal to the CR orbits of M. In the coordinates $(Z^0, u) \in \mathbb{C}^{N-r_3} \times \mathbb{C}^{r_3}$ the required parametrization has the form

$$H(Z^0, u) = \Gamma\big(\big(\partial^\alpha H(0, u)\big)_{|\alpha| \le k}, Z^0, u\big), \tag{6.1}$$

where k is a number depending only on the dimension N, Γ is a $\mathbb{C}^N$-valued holomorphic map defined in some neighborhood of $j^k H(0) \times \{0\}$ in the space $J^k(\mathbb{C}^N, \mathbb{C}^N) \times \mathbb{C}^{r_3}$, with $j^k H(0) = \big(\partial^\alpha H(0, 0)\big)_{|\alpha| \le k}$ and $J^k(\mathbb{C}^N, \mathbb{C}^N)$ denoting the space of k-jets at 0 of holomorphic maps from $\mathbb{C}^N$ to $\mathbb{C}^N$. Here it is important to note that Γ does not depend on the formal mapping H. Equality in (6.1) is in the sense of formal power series in Z^0 and u. The identity (6.1) is a simplified version of the statement of Theorem 10.1 in **[BRZ 2000]**.

The third step is to use (6.1) to obtain a system of holomorphic equations which must be satisfied by the formal series components of $\left(\partial^\alpha H(0,u)\right)_{|\alpha|\le k}$. To this system we apply Artin's approximation theorem **[A 1968]** to conclude that there exists a convergent solution for this system of equations. This yields the existence of a holomorphic map $\widehat{H} : (\mathbb{C}^N, 0) \to (\mathbb{C}^N, 0)$, which is the desired biholomorphic equivalence between $(M, 0)$ and $(M', 0)$. In fact, we can even choose $\widehat{H}$ in such a way that its Taylor series coincides with that of H up to a preassigned order.

7. General conditions for the convergence of formal equivalences

We shall give here a result about convergence of formal equivalences between two germs (M, p) and (M', p') for p and p' in general position, more general than that given in **[BER 1999b]** mentioned above. We restrict ourselves to the case where M and M' are generic submanifolds of $\mathbb{C}^N$, i.e., $r_1 = 0$. By making use of Theorem 2.5 we may assume that $(M, p) = (\widetilde{M} \times \mathbb{C}^{r_2}, 0)$ such that $\widetilde{M} \subset \mathbb{C}^{N-r_3}$ is as in Theorem 2.5 (but with $r_1 = 0$). Similarly we assume $(M', p') = (\widetilde{M}' \times \mathbb{C}^{r_2}, 0)$, with again $\widetilde{M}'$ as in Theorem 2.5, with the same integers r_2 and r_3. We shall assume (M, p) and $M', p')$ have this form for the remainder of this section.

Theorem 7.1 (**[BRZ 2000]**, Corollary 10.3). *Let (M, p) and (M', p') be as above. Then there exists an integer $\kappa \ge 0$ such that a formal equivalence H between (M, p) and (M', p') is convergent if and only if, for*

$$Z = (Z^0, Z^3, Z^2) \in \mathbb{C}^{N-r_2-r_3} \times \mathbb{C}^{r_3} \times \mathbb{C}^{r_2}, \quad H = (H^0, H^3, H^2),$$

both of the following conditions are satisfied.

(i) *All partial derivatives* $\frac{\partial^{|\alpha|}(H^0, H^3)}{\partial (Z^0)^\alpha}(0, Z^3, 0)$ *are convergent for* $|\alpha| \le \kappa$;

(ii) $H^2(Z)$ *is convergent.*

In fact, κ can be chosen to be $2(d+1)l$, where d is the codimension of $\widetilde{M}$ in $\mathbb{C}^{N-r_2}$ and l is chosen such that $\widetilde{M}$ is l-nondegenerate.

In the case where $r_2 = 0$, the proof of Theorem 7.1 follows immediately from the parametrization of all formal mappings given by (6.1), since in that case condition (i) of the theorem implies that the right hand side of (6.1) is convergent and hence so is the left hand side, $H(Z)$. For the general case where r_2 need not be 0, one also needs to use Proposition 5.3.

8. Real-algebraic submanifolds

A submanifold $M \subset \mathbb{C}^N$ is *real-algebraic* if it is contained in a real-algebraic subvariety of the same dimension. The basic example is given by the sphere $\sum_{j=1}^N |z_j|^2 = 1$. Most examples of real submanifolds included in this article are real-algebraic. The study of local biholomorphic maps sending pieces of spheres into each other goes back to Poincaré **[P 1907]** and Tanaka **[T 1962]**. Webster proved in **[W 1977]** that local biholomorphic maps sending open pieces of Levi-nondegenerate real-algebraic hypersurfaces into each other are complex-algebraic, i.e., their graphs are contained in complex-algebraic subvarieties of the same dimension. The algebraic properties of holomorphic maps sending one real-algebraic submanifold into another reveal the optimal nondegeneracy conditions for points in general position and have been intensively studied (see **[S 1991]**, **[H 1994]**, **[BR 1995]**, **[S 1995]**, **[Z 1995]**, **[BER 1996]**, **[SS 1996]**, **[Mi 1998]**, **[CMS 1999]**, **[Z 1999]**).

If M is a connected real-algebraic submanifold of $\mathbb{C}^N$, we say that a property holds for points in *general algebraic position* if it holds for all $p \in M$ outside a proper real-algebraic subvariety of M. Also, a stronger notion of equivalence, that of algebraic equivalence, can be naturally considered. Two germs of real-analytic submanifolds of $\mathbb{C}^N$, (M, p) and (M', p'), are said to be *algebraically equivalent* if there exists a biholomorphic equivalence between them which is complex-algebraic. Then the analogues of Propositions 2.1, 2.2, 2.3 and hence of Theorem 2.5 also hold in the category of real-algebraic submanifolds and algebraic equivalences for points in general algebraic position. The proof is based on the algebraic version of the implicit function theorem and other elementary properties (see e.g. **[BER 1999a]**, §5.4). In particular, the algebraic analogue of Proposition 2.1 follows from **[BER 1999a]**, Proposition 5.4.3 (d). The algebraic analogue of Proposition 2.2 can be obtained by repeating the proof of Proposition 3.1 in **[BRZ 2000]**. In contrast to this, the argument of the proof of Proposition 3.3 in **[BRZ 2000]** cannot be directly adapted to the algebraic case since the algebraic version of the Frobenius theorem does not hold. The algebraic analogue of Proposition 2.3 was shown in **[BER 1996]** (Lemma 3.4.1) by using the Segre sets rather than the Frobenius theorem. In particular, the CR orbits of real-algebraic CR submanifolds are algebraic (**[BER 1996]**, Corollary 2.2.5), whereas the orbits of single vector fields in T^cM (the real parts of (0, 1) vector fields) need not be algebraic.

Example 8.1. Consider the real-algebraic hypersurface $M \subset \mathbb{C}^2$ given in real coordinates by $y_2 = x_2 y_1$ where $(z_1, z_2) = (x_1 + iy_1, x_2 + iy_2)$. The sections of the complex tangent subbundle T^cM are spanned at each point by the vector fields

$$X = x_2 \frac{\partial}{\partial x_2} + \frac{\partial}{\partial x_1} - y_1 \frac{\partial}{\partial y_1}, \quad Y = JX_1 = x_2 \frac{\partial}{\partial y_2} + \frac{\partial}{\partial y_1} + y_1 \frac{\partial}{\partial x_1}.$$

The integral curve C for X through $p_0 = (x_1^0, x_2^0, y_1^0, y_2^0) \in M$ is given by

$$x_2 = x_2^0 e^{x_1 - x_1^0}, \quad y_1 = y_1^0 e^{x_1^0 - x_1}, \quad y_2 = y_2^0.$$

Hence C is not algebraic if $x_2^0 \neq 0$ or $y_1^0 \neq 0$. (In contrast, the orbits of Y are all algebraic.) However, the CR orbit of any point $p_0 = (x_1^0, x_2^0, y_1^0, y_2^0) \in M$ with $x_2^0 \neq 0$ is (M, p_0), while when $x_2^0 = 0$ (and hence $y_2^0 = 0$) the CR orbit is $(\mathbb{C} \times \{0\}, p_0)$. Thus the algebraicity of these CR orbits cannot be proved by showing the algebraicity of the integral curves of the basis of sections of $T^c M$ given by X, Y.

9. Algebraic equivalence for real-algebraic submanifolds

The following strengthens Theorem 6.1 in the case of real-algebraic submanifolds.

Theorem 9.1. *Let $M \subset \mathbb{C}^N$ be a connected real-algebraic submanifold. Then for any $p \in M$ in general algebraic position and any germ (M', p') of a real-algebraic submanifold in $\mathbb{C}^N$, the equivalent conditions* (i)–(iv) *of Theorem* 6.1 *are also equivalent to*

(v) *(M, p) and (M', p') are algebraically equivalent.*

The proof of Theorem 9.1 can be obtained by adapting the proof of Theorem 6.1 given in **[BRZ 2000]** to the algebraic case. It is sufficient to prove that (iv) implies (v). As in the proof of Theorem 6.1 the first step is to reduce the general situation by a complex-algebraic change of coordinates to the case where $p = p' = 0$, $M = \widetilde{M}$ and $M' = \widetilde{M}'$ with $\widetilde{M}$ and $\widetilde{M}'$ as in Theorem 2.5, or rather its real-algebraic analogue mentioned above. Here again, $r_1 = r_2 = 0$. The second step is to obtain an algebraic parametrization of all biholomorphic equivalences H between $(M, 0)$ and $(M', 0)$ by their jets along the linear subspace $C = \{0\} \times \mathbb{C}^{r_3}$. This parametrization takes the form (6.1) with Γ a map which is not only holomorphic but also complex-algebraic, defined in some neighborhood of $j^k H(0) \times \{0\}$ in the jet space $J^k(\mathbb{C}^N, \mathbb{C}^N) \times \mathbb{C}^{r_3}$. Note that in this case the right hand side of (6.1) is a convergent power series in (Z_0, u). If $u \mapsto \big(\partial^\alpha H(0, u)\big)_{|\alpha| \leq k}$ is complex algebraic, then the algebraicity of H follows immediately from the analogue of (6.1). Hence in the special case when M is of finite type, i.e., $r_3 = 0$, H is parametrized by the single jet $j^k H(0) = \big(\partial^\alpha H(0)\big)_{|\alpha| \leq k}$ so that the biholomorphic equivalence H is algebraic itself. If $r_3 > 0$, the jet $\big(\partial^\alpha H(0, u)\big)_{|\alpha| \leq k}$ and hence H need not be algebraic. Then we have to modify $\big(\partial^\alpha H(0, u)\big)_{|\alpha| \leq k}$ to obtain an algebraic equivalence. This is done by using an algebraic version of Artin's approximation theorem **[A 1969]**. Here again we can choose the Taylor series of the algebraic equivalence to coincide with that of H up to a preassigned order.

Example 9.1. Let $M_0 \subset \mathbb{C}^{N-1}$ be an arbitrary finitely nondegenerate real-algebraic submanifold through 0 and set $M = M' := M_0 \times \mathbb{R} \subset \mathbb{C}^N$. Then both M and M' are finitely nondegenerate of infinite type and any biholomorphic map $H(z, w) := (z, \phi(w))$ with $\phi\colon (\mathbb{C}, 0) \to (\mathbb{C}, 0)$ sending $\mathbb{R}$ into itself is a biholomorphic equivalence between $(M, 0)$ and $(M', 0)$. It is clear that if ϕ is not algebraic, then H is not algebraic. On the other hand, the identity mapping is an algebraic equivalence between $(M, 0)$, and $(M', 0)$, and there are many others.

References

[A 1968] Artin, M., On the solutions of analytic equations, Invent. Math. 5 (1968), 277–291.

[A 1969] Artin, M., Algebraic approximation of structures over complete local rings, Inst. Hautes Etudes Sci. Publ. Math. 36 (1969), 23–58.

[BR 1995] Baouendi, M. S., Rothschild, L. P., Mappings of real algebraic hypersurfaces, J. Amer. Math. Soc. 8 (1995), 997–1015.

[BER 1996] Baouendi, M. S., Ebenfelt, P., Rothschild, L. P., Algebraicity of holomorphic mappings between real algebraic sets in $\mathbb{C}^n$, Acta Math. 177 (1996), 225–273.

[BER 1999a] Baouendi, M. S., Ebenfelt, P., and Rothschild, L. P., Real Submanifolds in Complex Space and Their Mappings, Princeton Math. Ser. 47, Princeton University Press, 1999.

[BER 1999b] Baouendi, M. S., Ebenfelt, P., and Rothschild, L. P., Rational dependence of smooth and analytic CR mappings on their jets, Math. Ann. 315 (1999), 205–249.

[BER 2000] Baouendi, M. S., Ebenfelt, P., and Rothschild, L. P., Local geometric properties of real submanifolds in complex space, Bull. Amer. Math. Soc. 37 (2000), 309–336.

[BHR 1996] Baouendi, M. S., Huang, X., and Rothschild, L. P., Regularity of CR mappings between algebraic hypersurfaces, Invent. Math. 125 (1996), 13–36.

[BRZ 2000] Baouendi, M. S., Rothschild, L. P., and Zaitsev, D.: Equivalences of real submanifolds in complex space, preprint, 2000, `http://xxx.lanl.gov/abs/math.CV/0002186`.

[BG 1977] Bloom, T., Graham, I., On type conditions for generic real submanifolds of $\mathbb{C}^n$, Invent. Math. 40 (1977), 217–243.

[Bo 1991] Boggess, A., CR Manifolds and the Tangential Cauchy–Riemann Complex, Stud. Adv. Math., CRC Press, Boca Raton–Ann Arbor–Boston–London, 1991.

[Ca 1932a] Cartan, E., Sur la géométrie pseudo-conforme des hypersurfaces de deux variables complexes, I, Ann. Math. Pura Appl. 11 (1932), 17–90; Œuvres complètes, Part. II, Vol. 2, Gauthier-Villars, 1952, 1231–1304.

[Ca 1932b] Cartan, E., Sur la géométrie pseudo-conforme des hypersurfaces de deux variables complexes, II, Ann. Scuola Norm. Sup. Pisa 1 (1932), 333–354; Œuvres complètes, Part. III, Vol. 2, Gauthier-Villars, 1952, 1217–1238.

[Ca 1937] Cartan, E., Les problèmes d'équivalence, Séminaire de Math., exposé D, 11 janvier 1937; Selecta, 113–136; Œuvres complètes, Part. II, Vol. 2, Gauthier-Villars, 1952, 1311–1335.

[CM 1974] Chern, S. S, Moser, J. K., Real hypersurfaces in complex manifolds, Acta Math. 133 (1974), 219–271.

[Ch 1991] Chirka, E. M., Introduction to the geometry of CR-manifolds, Uspekhi Mat. Nauk 46 (1) (1991), 81–164 = Russ. Math. Surveys 46 (1) (1991), 95–197.

[CMS 1999] Coupet, B., Meylan, F., Sukhov, A., Holomorphic maps of algebraic CR manifolds, Internat. Math. Res. Notices 1 (1999), 1–29.

[E 1998] Ebenfelt, P., New invariant tensors in CR structures and a normal form for real hypersurfaces at a generic Levi degeneracy, J. Differential Geom. 50 (1998), 207–247.

[E 2000] Ebenfelt, P., Finite jet determination of holomorphic mappings at the boundary, preprint, 2000, Asian J. Math., to appear; `http://xxx.lanl.gov/abs/math.CV/0001116`.

[F 1977] Freeman, M, Local biholomorphic straightening of real submanifolds, Ann. Math. 106 (1977), 319–352.

[H 1994] Huang, X., On the mapping problem for algebraic real hypersurfaces in the complex spaces of different dimensions, Ann. Inst. Fourier 44 (1994), 433–463.

[J 1990] Jacobowitz, H., An introduction to CR structures, Math. Surveys Monogr. 32, Amer. Math. Soc., Providence, RI, 1990.

[K 1972] Kohn, J. J., Boundary behavior of $\bar{\partial}$ on weakly pseudo-convex manifolds of dimension two, J. Differential Geom. 6 (1972), 523–542,.

[Mi 1998] Mir, N., Germs of holomorphic mappings between real algebraic hypersurfaces, Ann. Inst. Fourier 48 (1998), 1025–1043.

[MW 1983] Moser, J. and Webster, S. M., Normal forms for real surfaces in $\mathbb{C}^2$ near complex tangents and hyperbolic surface transformations, Acta Math. 150 (1983), 255–296.

[P 1907] Poincaré, H., Les fonctions analytiques de deux variables et la représentation conforme, Rend. Circ. Mat. Palermo 23 (1907), 185–220.

[SS 1996] Sharipov, R; Sukhov, A., On CR mappings between algebraic Cauchy–Riemann manifolds and separate algebraicity, Trans. Amer. Math. Soc. 348 (1996), 767–780.

[S 1991] Sukhov, A. B., On the algebraicity of complex analytic sets, Mat. Sb. 182 (1991), 1669–1677 = Math. USSR-Sb. 74 (1993), 419–426.

[S 1995] Sukhov, A. B., A remark on the algebraicity of complex manifolds, Mat. Zametki 57 (2) (1995), 315–316 = Math. Notes 57 (1–2) (1995), 225–226.

[T 1962] Tanaka, N., On the pseudo-conformal geometry of hypersurfaces of the space of n complex variables, J. Math. Soc. Japan 14 (1962), 397–429.

[W 1977] Webster, S., On the mapping problem for algebraic real hypersurfaces, Invent. Math. 43 (1977), 53–68.

[W 1995] Webster, S., The holomorphic contact geometry of a real hypersurface, in: Modern Methods in Complex Analysis (Bloom, Th. et al., eds.), Ann. of Math. Stud. 137, Princeton University Press, 1995, 327–342.

[Z 1995] Zaitsev, D., On the automorphism groups of algebraic bounded domains, Math. Ann. 302 (1995), 105–129.

[Z 1999] Zaitsev, D., Algebraicity of local holomorphisms between real-algebraic submanifolds of complex spaces, Acta Math. 183 (1999), 273–305.

Department of Mathematics
University of California at San Diego
La Jolla, CA 92093-0112, U.S.A.
sbaouendi@ucsd.edu
lrothschild@ucsd.edu

Mathematisches Institut
Universität Tübingen
Auf der Morgenstelle 10
72076 Tübingen, Germany
dmitri.zaitsev@uni-tuebingen.de

Holomorphic motion of circles through affine bundles

David E. Barrett *

1. Introduction

The pivotal topic of this paper is the study of Levi-flat real hypersurfaces S with circular fibers in a rank 1 affine bundle A over a Riemann surface X. (To say that S is Levi-flat is to say that S admits a foliation by Riemann surfaces; equivalently, in the language of **[SuTh]**, S may be said to prescribe a holomorphic motion of circles through A.)

After setting notation and terminology in §2 we proceed in §3 to examine the Levi-form of a general real hypersurface with circular fibers, emphasizing the connection with curvature considerations.

In §4 we focus on the Levi-flat case. In Theorems 5 and 6 we construct moduli spaces for Levi-flat S attached to a fixed underlying line bundle L in the compact and non-compact cases, respectively. In particular, when X is compact we show that the existence of a Levi-flat S implies that $0 \leq \deg L \leq 2\,\mathrm{genus}(X) - 2$. (The bound is sharp.)

Theorem 7 in §7 states that when S is Levi-flat, the Levi-foliation on S extends to a holomorphic foliation of the $\mathbb{CP}^1$ bundle obtained from A by compactifying the fibers. In the general case, the extended foliation in constructed by looking for holomorphic sections of A whose distance from the center is harmonic with respect to the appropriate metric. In §7 we show that this construction produces a foliation even in some cases where S "disappears into the recomplexification of A."

§6 looks at general holomorphic foliations (transverse to fibers) of compactified rank 1 affine bundles; in particular, it is shown that such foliations are classified up to equivalence by a "Schwarzian derivative" and a "curvature function." An Addendum to Theorem 7 shows how to recognize when such a foliation arises from a Levi-flat hypersurface.

The remaining sections contain postponed proofs.

*Supported in part by the National Science Foundation.

2. Notation and terminology

2.1. Affine bundles. Let L be a holomorphic line bundle over a Riemann surface X; L can be defined by local trivializations with transition functions of the form

$$(z, w) \mapsto \left(\phi_{\alpha,\beta}(z), \chi_{\alpha,\beta}(z) \cdot w\right). \tag{2.1}$$

An affine bundle A over X associated to L can be defined by local trivializations with transition functions of the form

$$(z, w) \mapsto \left(\phi_{\alpha,\beta}(z), \chi_{\alpha,\beta}(z) \cdot w + \sigma_{\alpha,\beta}(z)\right)$$

satisfying the appropriate cocycle condition. Over each point $\zeta \in X$ we have a well-defined subtraction operation $A_\zeta \times A_\zeta \to L_\zeta$ defined in local bundle coordinates by

$$((z(\zeta), w_1), (z(\zeta), w_2) \mapsto (z(\zeta), w_1 - w_2).$$

We will use the term *L-shear* to refer to a biholomorphic map between affine bundles A and A' over X associated to L taking each fiber A_ζ to the corresponding fiber A'_ζ and preserving the subtraction operation.

Let γ be a smooth section of an affine bundle A associated to L. Then $\bar\partial\gamma$ defines a section of $L \otimes T^{*(0,1)}(X)$.

The following result follows easily from the definitions.

Proposition 1. *Let A_1 and A_2 be affine bundles over X associated to a fixed line bundle L, and let γ_j be a smooth section of A_j. Then the following conditions are equivalent:*

(1) *there is an L-shear from A_1 to A_2 carrying γ_1 to γ_2;*

(2) $\bar\partial\gamma_1 = \bar\partial\gamma_2$.

Conversely, let ω be a smooth section of $L \otimes T^{*(0,1)}(X)$. Then using a system of local solutions of $\bar\partial u = \omega$ we may construct an affine bundle A associated to L and a section γ of A satisfying $\bar\partial\gamma = \omega$. Alternatively we may accomplish the same end by taking the total space of A to be the total space of L equipped with the unique complex structure J_ω satisfying:

- J_ω coincides with the standard structure J_0 on vectors tangent to fibers;
- a (local) section γ of L is J_ω-holomorphic if and only if it solves $\bar\partial\gamma = -\omega$ (with respect to the standard structure J_0).

Using local coordinates (z, w) coming from a local trivialization of L we find that the (1,0) tangent vector fields for the structure J_ω are spanned by $\frac{\partial}{\partial w}$ and

$\frac{\partial}{\partial z} - \overline{\langle \omega, \frac{\partial}{\partial \bar{z}} \rangle} \frac{\partial}{\partial \bar{w}}$. The integrability of J_ω can be checked directly; alternately we may note that a solution of $\bar{\partial}\gamma = -\omega$ on a open set U induces a biholomorphic map

$$\begin{aligned} \left(L\big|_U, J_0\right) &\to \left(L\big|_U, J_\omega\right) \\ (z, w) &\mapsto (z, w + \gamma(z)) \,. \end{aligned}$$

It follows that (L, J_ω) is an affine bundle over X associated to L; since the induced Cauchy–Riemann operator $\bar{\partial}_\omega$ satisfies $\bar{\partial}_\omega = \bar{\partial}_0 + \omega$ we find that the zero section of L provides a distinguished smooth section of (L, J_ω) satisfying $\bar{\partial}_\omega 0 = \omega$.

2.2. Bundle metrics; hypersurfaces with circular fibers. Let A be an affine bundle over X associated to the line bundle L and let γ be a smooth section of A. Suppose now that the line bundle L is equipped with a Hermitian metric h. Then we may consider the real hypersurface $S = S_{\gamma,h} \subset A$ whose fiber over $\zeta \in X$ is the circle centered at $\gamma(\zeta)$ with unit radius with respect to h. Using bundle coordinates (z, w) and writing $h = e^{u(z)}\,|dw|$ we find that $S_{\gamma,h}$ is given by the equation $|w - \gamma(z)| = e^{-u(z)}$.

Proposition 2. *The map*

$$\left[S_{\gamma,h}\right] \mapsto \left(\bar{\partial}\gamma, h\right)$$

is a bijection between

$$\frac{\{\textit{real hypersurfaces with circular fibers in affine bundles associated to } L\}}{\{L\textit{-shears}\}}$$

and

$$\{\textit{smooth sections of } L \otimes T^{*(0,1)}(X)\} \times \{\textit{Hermitian metrics on } L\}.$$

In particular, $S_{\gamma,h}$ *is equivalent up to* L*-shears to the unit circle bundle* $\Sigma_{\bar{\partial}\gamma,h} \subset \left(L, J_{\bar{\partial}\gamma}\right)$.

3. Levi-form computations

Remarks on terminology. If L is a (not necessarily holomorphic) line bundle given by transition functions (2.1), we let $|L|$ denote the line bundle given by transition functions

$$(z, w) \mapsto \left(\phi_{\alpha,\beta}(z), |\chi_{\alpha,\beta}(z)| \cdot w\right).$$

We will frequently find it convenient to identify $|L|^2$ with $L \otimes \overline{L}$.

A metric on L may be viewed as a section of $|L|^{-1}$.

Proposition 3. *Let L be a line bundle over a Riemann surface X, let ω be a smooth section of $L \otimes T^{*(0,1)}(X)$, and let h be a Hermitian metric on L. Then the unit circle bundle $\Sigma_{\omega,h}$ in (L, J_ω) is*

$$\left\{\begin{matrix} \textit{strictly pseudoconcave} \\ \textit{pseudoconcave} \\ \textit{Levi-flat} \\ \textit{pseudoconvex} \\ \textit{strictly pseudoconvex} \end{matrix}\right\}$$

if and only if

$$\left\{\begin{matrix} \frac{i}{2}\Theta > ih^2\omega \wedge \overline{\omega} + h^{-1}\left|\partial\left(h^2\omega\right)\right| \\ \frac{i}{2}\Theta \geq ih^2\omega \wedge \overline{\omega} + h^{-1}\left|\partial\left(h^2\omega\right)\right| \\ \frac{1}{2}\Theta = h^2\omega \wedge \overline{\omega},\ \partial\left(h^2\omega\right) = 0 \\ \frac{i}{2}\Theta \leq ih^2\omega \wedge \overline{\omega} - h^{-1}\left|\partial\left(h^2\omega\right)\right| \\ \frac{i}{2}\Theta < ih^2\omega \wedge \overline{\omega} - h^{-1}\left|\partial\left(h^2\omega\right)\right| \end{matrix}\right\};$$

here $\Theta = -2\partial\bar{\partial}\log h$ is the curvature $(1,1)$-form for the metric h, and the inequalities are taken with respect to the standard orientation on X.

To further explain the above equations, note that

- $h^2\omega \wedge \overline{\omega}$ is a section of $|L|^{-2} \otimes \left(L \otimes T^{*(0,1)}\right) \otimes \left(\overline{L} \otimes T^{*(1,0)}\right) = T^{*(1,1)}$;
- $h^2\omega$ is a section of the anti-holomorphic bundle $|L|^{-2} \otimes L \otimes T^{*(0,1)} = \overline{L}^{-1} \otimes T^{*(0,1)}$;
- $\partial\left(h^2\omega\right)$ is a section of $\overline{L}^{-1} \otimes T^{*(1,1)}$;
- $\left|\partial\left(h^2\omega\right)\right|$ is a section of $|L|^{-1} \otimes T^{*(1,1)}$;
- $h^{-1}\left|\partial\left(h^2\omega\right)\right|$ is a section of $T^{*(1,1)}$.

By Proposition 2, the results of Proposition 3 also describe the pseudoconvexity properties of hypersurfaces $S_{\gamma,h}$ with $\bar{\partial}\gamma = \omega$. Passing to local coordinates as in §2.2 this translates to the statement that the hypersurface $|w - \gamma(z)| = e^{-u(z)}$ is

$$\left\{\begin{matrix} \text{strictly pseudoconcave} \\ \text{pseudoconcave} \\ \text{Levi-flat} \\ \text{pseudoconvex} \\ \text{strictly pseudoconvex} \end{matrix}\right\} \text{ if and only if } \left\{\begin{matrix} -u_{z\bar{z}} > e^{2u}\left|\gamma_{\bar{z}}\right|^2 + e^u\left|\gamma_{z\bar{z}} + 2u_z\gamma_{\bar{z}}\right| \\ -u_{z\bar{z}} \geq e^{2u}\left|\gamma_{\bar{z}}\right|^2 + e^u\left|\gamma_{z\bar{z}} + 2u_z\gamma_{\bar{z}}\right| \\ -u_{z\bar{z}} = e^{2u}\left|\gamma_{\bar{z}}\right|^2,\ \gamma_{z\bar{z}} + 2u_z\gamma_{\bar{z}} = 0 \\ -u_{z\bar{z}} \leq e^{2u}\left|\gamma_{\bar{z}}\right|^2 - e^u\left|\gamma_{z\bar{z}} + 2u_z\gamma_{\bar{z}}\right| \\ -u_{z\bar{z}} < e^{2u}\left|\gamma_{\bar{z}}\right|^2 - e^u\left|\gamma_{z\bar{z}} + 2u_z\gamma_{\bar{z}}\right| \end{matrix}\right\}.$$

For a proof in this framework see [**Ber**, Prop. 2.3] and the references cited there.

For $\omega \equiv 0$ Proposition 3 reduces to the following (quite classical) result.

Corollary 4. *Let L be a line bundle over a Riemann surface X equipped with a Hermitian metric h. Then the unit circle bundle $\Sigma_{0,h}$ in L is*

$$\left\{\begin{matrix} \textit{strictly pseudoconcave} \\ \textit{pseudoconcave} \\ \textit{Levi-flat} \\ \textit{pseudoconvex} \\ \textit{strictly pseudoconvex} \end{matrix}\right\}$$

if and only if the curvature form Θ satisfies

$$\left\{\begin{matrix} i\Theta > 0 \\ i\Theta \geq 0 \\ \Theta = 0 \\ i\Theta \leq 0 \\ i\Theta < 0 \end{matrix}\right\}.$$

For vector bundles of higher dimension, curvature conditions for Hermitian and Finsler metrics are related to the theory of interpolation of norms **[Roc]**.

4. The Levi-flat case

Recall that a real hypersurface in a complex manifold is said to be *Levi-flat* if its Levi-form vanishes identically. A real hypersurface is Levi-flat if and only if it admits a (uniquely-determined) codimension-one foliation with complex leaves **[Kra**, p. 308**]**.

In the situation of Proposition 3, if $\Sigma_{\omega,h} \subset (L, J_\omega)$ is Levi-flat then $h^2\overline{\omega}$ is a holomorphic section of $L^{-1} \otimes T^{*(1,0)}(X)$.

If $h^2\overline{\omega} \equiv 0$ then by Proposition 1 we may take $A = L$; also the curvature Θ of the metric h vanishes identically so that h is flat. In the case where X is compact we thus have $\deg L = 0$ **[GrHa**, §1.1**]**. (Recall that every degree 0 line bundle admits a flat metric, unique up to scalar multiples **[GrHa**, §1.2**]**.)

To analyze the case where $h^2\overline{\omega}$ does not vanish identically, note that $2h^{-1}\left|h^2\overline{\omega}\right| = 2h|\omega|$ is a non-negative section of $\left|T^{*(1,0)}(X)\right|$; it may be viewed as a conformal metric on X with a so-called conical singularity of total angle $2\pi(j+1)$ at any point where $h^2\overline{\omega}$ has a zero of order j (see for example **[HuTr**, §2**]**). The corresponding area form is $2ih^2\omega \wedge \overline{\omega}$, and away from the degeneracies the scalar curvature is given by

$$\begin{aligned} -\frac{2i\partial\bar{\partial}\log\left(2h^{-1}\left|h^2\overline{\omega}\right|\right)}{2ih^2\omega \wedge \overline{\omega}} &= \frac{2i\partial\bar{\partial}\log h}{2ih^2\omega \wedge \overline{\omega}} \\ &= -\frac{\Phi/2}{h^2\omega \wedge \overline{\omega}} \\ &= -1. \end{aligned}$$

To take proper account of the degeneracies we may compute the Gauss–Bonnet form in the sense of distributions:

$$-2i\partial\bar\partial \log\left(2h^{-1}\left|h^2\overline{\omega}\right|\right)$$
$$= -2ih^2\omega\wedge\overline{\omega} - 2\pi \sum_{\{\zeta\in X:\omega(\zeta)=0\}} \left(\text{order of vanishing of } h^2\overline{\omega} \text{ at } \zeta\right)\cdot\delta_\zeta,$$

where δ_ζ denotes a unit point mass at ζ. (See for example **[Bar**, Lemma 11**]**.)

In the case where X is compact, invocation of the Gauss–Bonnet theorem yields

$$\begin{aligned}
&4\pi(1-\text{genus}(X))\\
&= -\int_X 2i\partial\bar\partial\log\left(2h^{-1}\left|h^2\overline{\omega}\right|\right)\\
&= -\int_X 2ih^2\omega\wedge\overline{\omega} - 2\pi\sum_{\{\zeta\in X:\omega(\zeta)=0\}}\left(\text{order of vanishing of } h^2\overline{\omega}\text{ at }\zeta\right)\\
&= -\int_X 2ih^2\omega\wedge\overline{\omega} - 2\pi\deg\left(L^{-1}\otimes T^{*(1,0)}(X)\right)\\
&= -\int_X 2ih^2\omega\wedge\overline{\omega} - 2\pi\,(2(\text{genus}(X)-1)-\deg L)
\end{aligned}$$

so that

$$\deg L = \frac{1}{\pi}\int_X ih^2\omega\wedge\overline{\omega} > 0.$$

Conversely, suppose that X is compact and that $\deg L > 0$. Then for any non-trivial holomorphic section f of $L^{-1}\otimes T^{*(1,0)}(X)$ the results of **[HuTr**, Thm. B**]** allow us to construct (uniquely) a conformal metric $\tilde h$ on X of curvature -1 with conical singularities of total angle $2\pi(j+1)$ at points where f has a zero of order j. Setting $h = 2\tilde h^{-1}|f|$, $\omega = \frac{\tilde h^2}{4f}$ we find that $\Sigma_{\omega,h}\subset(L, J_\omega)$ is Levi-flat.

Let $\text{LFC}(X, L)$ denote the space

$$\frac{\{\text{Levi-flat hypersurfaces with circular fibers in affine bundles associated to } L\}}{\{L\text{-shears}\}}.$$

Then we may sum up the preceding discussion as follows.

Theorem 5. *Let X be a compact Riemann surface and let L be a holomorphic line bundle on X.*

If $\deg L < 0$ *then* $\text{LFC}(X, L) = \emptyset$.

If $\deg L = 0$ *then the map* $\left[S_{\gamma,h}\right]\mapsto h$ *is a bijection between* $\text{LFC}(X, L)$ *and the one-dimensional space of flat metrics on* L.

If $\deg L > 0$ *then the map* $\left[S_{\gamma,h}\right]\mapsto h^2\partial\overline{\gamma}$ *is a bijection between* $\text{LFC}(X, L)$ *and the space of non-trivial holomorphic sections of* $L^{-1}\otimes T^{*(1,0)}(X)$.

Note that if $\deg L > 2\,\mathrm{genus}(X) - 2$ then $\deg\left(L^{-1} \otimes T^{*(1,0)}(X)\right) < 0$ so that there are no non-trivial holomorphic sections of $L^{-1} \otimes T^{*(1,0)}(X)$; thus in this case we again have $\mathrm{LFC}(X, L) = \emptyset$.

Note also that for $\deg L > 0$ we never have $A = L$, else we would have

$$0 = -\int_X \bar{\partial}\left(h^2\, \partial\overline{\gamma}\right)\gamma = \int_X h^2 \left|\bar{\partial}\gamma\right|^2,$$

forcing $h^2 \partial\overline{\gamma} \equiv 0$.

To treat the case of non-compact X we will get a simpler-to-state result by working modulo not just shears but arbitrary fiber-preserving biholomorphic maps – following terminology in dynamics **[AnLe]** we will refer to such maps as *overshears*.

Theorem 6. *If A is a rank 1 affine bundle over a noncompact Riemann surface X then the map $\left[S_{\gamma,h}\right] \mapsto 2h\left|\bar{\partial}\gamma\right|$ is a bijection from*

$$\frac{\{\textit{Levi-flat } S_{\gamma,h} \subset A\colon \gamma \textit{ not holomorphic}\}}{\{\textit{overshears}\}}$$

to the space

$$\{\textit{conformal metrics on } X \textit{ of curvature } -1 \textit{ with all total angles} \in 2\pi\mathbb{N}\}.$$

Proof. Recall from Proposition 2 that S_{γ_1,h_1} and S_{γ_2,h_2} are equivalent modulo shears if and only if $h_1 = h_2$ and $\bar{\partial}\gamma_1 = \bar{\partial}\gamma_2$. Similarly, it is straightforward to check that S_{γ_1,h_1} and S_{γ_2,h_2} are equivalent modulo overshears if and only if there is $g : X \to \mathbb{C}\setminus\{0\}$ holomorphic with $h_1 = |g|h_2$, $\bar{\partial}\gamma_1 = g^{-1}\bar{\partial}\gamma_2$; the latter condition implies that $2h_1|\bar{\partial}\gamma_1| = 2h_2|\bar{\partial}\gamma_2|$, showing that our map is well-defined.

To check injectivity, note if S_{γ_1,h_1} and S_{γ_2,h_2} are Levi-flat then equality of $2h_1|\bar{\partial}\gamma_1|$ and $2h_2|\bar{\partial}\gamma_2|$ forces the holomorphic sections $h_1^2\,\partial\overline{\gamma}_1$ and $h_2^2\,\partial\overline{\gamma}_2$ to have the same zeros (counting multiplicities) so that $h_1^2\,\partial\overline{\gamma}_1 = g h_2^2\,\partial\overline{\gamma}_2$ for some holomorphic $g : X \to \mathbb{C}\setminus\{0\}$. Thus

$$h_1 = \frac{2\left|h_1^2\,\partial\overline{\gamma}_1\right|}{2h_1\left|\bar{\partial}\gamma_1\right|} = |g|h_2 \qquad \text{and} \qquad \bar{\partial}\gamma_1 = \frac{\left(2h_1\left|\bar{\partial}\gamma_1\right|\right)^2}{4h_1^2\,\partial\overline{\gamma}_1} = g^{-1}\,\bar{\partial}\gamma_2,$$

showing that S_{γ_1,h_1} and S_{γ_2,h_2} are equivalent modulo overshears, as required.

To prove surjectivity, let h be a conformal metric on X of curvature -1 with all total angles $\in 2\pi\mathbb{N}$. By the Weierstraß Product Theorem **[For**, Thm. 26.5**]** and the triviality of $L^{-1} \otimes T^{*(1,0)}(X)$ **[For**, Thm. 30.3**]** we may pick a holomorphic section f of $L^{-1} \otimes T^{*(1,0)}(X)$ so that f vanishes to order j at ζ if and only if $\widetilde{h}$ has total angle $2\pi(j+1)$ at ζ. Let $h = 2\widetilde{h}^{-1}|f|$, $\omega = \frac{\widetilde{h}^2}{4f}$. Then $\Sigma_{\omega,h} \subset (L, J_\omega)$ is Levi-flat, and our mapping takes $[\Sigma_{\omega_h}]$ to $\widetilde{h}$, as required. □

Turning to the flat case, a standard argument shows that when X is non-compact, $S_{\gamma,h}$ with γ holomorphic are classified up to overshears by the associated monodromy homomorphism $\pi_1(X) \to S^1$ **[CaLN**, Chap. V].

We close this section with consideration of the special case where $L = T^{*(1,0)}(X)$ (with X not necessarily compact) and the (1,1)-form ω is positive. Then $h^2\overline{\omega}$ is both holomorphic and positive, hence equal to a constant $C/2$; it follows that h is $\sqrt{C}$ times the metric on $T^{*(1,0)}(X)$ induced by the conformal metric on X with area form ω. Moreover, $2h|\omega| = Ch^{-1}$ has curvature -1, so the conformal metric on X with area form ω has curvature $-C$.

5. Extension of Levi foliations

If A is an affine bundle over X we will denote by $\widehat{A}$ the $\mathbb{CP}^1 = \widehat{\mathbb{C}} = \mathbb{C} \cup \{\infty\}$ bundle over X obtained by adding a point at infinity to each fiber A_ζ.

Theorem 7. *Let $S = S_{\gamma,h}$ be a Levi-flat hypersurface with circular fibers in an affine bundle A over a Riemann surface X. Then the Levi-foliation of S extends uniquely to a holomorphic foliation of $\widehat{A}$.*

The extended foliation $\mathcal{F}_S$ is transverse to the fibers $\widehat{A}_\zeta$.

If the corresponding line bundle metric h is flat then $\mathcal{F}_S$ is described by the condition

(CM) *the graph of a local holomorphic section ν of A lies in a leaf if and only if $\|\nu - \gamma\|$ is constant.*

If the corresponding line bundle metric h is not flat then $\mathcal{F}_S$ is described by the condition

(LHM) *the graph of a local holomorphic section ν of A lies in a leaf if and only if $\log \|\nu - \gamma\|$ is harmonic on $X \setminus (\nu - \gamma)^{-1}(0)$.*

Theorem 7 will be proved in §9.

6. Foliations of compactified affine bundles

6.1. Residues. Let A be an affine bundle over X associated to a line bundle L and let $\mathcal{F}$ be a holomorphic foliation of $\widehat{A}$ tranvserse to fibers $\widehat{A}_\zeta$. For $\zeta \in X$ let ν_ζ be the unique (germ of a) meromorphic section of A with graph contained in a leaf of $\mathcal{F}$ satisfying $\nu_\zeta(\zeta) = \infty$.

If ν_ζ has a simple pole at ζ then the residue $\operatorname{Res}_\zeta \nu_\zeta$ defines an element of $T_\zeta^{(1,0)} \otimes L_\zeta$. (To be specific, we may choose a small loop C_ζ about ζ and a local holomorphic section γ of A; we then define a functional Υ_ζ on $T_\zeta^{*(1,0)} \otimes L_\zeta^{-1}$ by the formula

$$\Upsilon_\zeta(\omega(\zeta)) = \frac{1}{2\pi i} \int_{C_\zeta} (\nu_\zeta - \gamma) \cdot \omega$$

for ω a local holomorphic section of $T^{*(1,0)}(X) \otimes L^{-1}$. Υ_ζ is clearly linear, and it is easy to check that the right hand side depends only on $\omega(\zeta)$ and in particular does not depend on the choice of C_ζ or γ. Then we can define $\operatorname{Res}_\zeta \nu_\zeta$ to be the element of $T_\zeta^{(1,0)}(X) \otimes L_\zeta$ representing Υ_ζ.)

We may define a section $\kappa_{\mathcal{F}}$ of $T^{*(1,0)}(X) \otimes L^{-1}$ by setting

$$\kappa_{\mathcal{F}}(\zeta) = -2i\left(\operatorname{Res}_\zeta \nu_\zeta\right)^{-1}$$

when ν_ζ has a simple pole at ζ and $\kappa_{\mathcal{F}}(\zeta) = 0$ when ν_ζ has a multiple pole at ζ.

Proposition 8. *For $\mathcal{F}$ as above, $\kappa_{\mathcal{F}}$ is a holomorphic section of $T^{*(1,0)}(X) \otimes L^{-1}$.*

Proof. Since the fibers of $\widehat{A}$ are compact, the transversality hypothesis guarantees that $\mathcal{F}$ is locally equivalent to a product foliation on $X \times \widehat{\mathbb{C}}$ **[CaLN**, Chap. V]. Thus, choosing bundle coordinates (z, w) for the restriction of A to a small open set $U \subset X$ we find that there are holomorphic functions $a(z), b(z), c(z), d(z)$ with $ad - bc \equiv 1$ so that leaves of $\mathcal{F}$ are given by equations of the form

$$\frac{a(z)w + b(z)}{c(z)w + d(z)} = C \tag{6.1}$$

or

$$w = \frac{d(z)C - b(z)}{-c(z)C + a(z)}, \tag{6.2}$$

C constant. To get $w = \infty$ when $z = \zeta$ we must set $C = C_\zeta = a(\zeta)/c(\zeta)$ so that

$$\nu_\zeta(z) = \frac{a(\zeta)d(z) - b(z)c(\zeta)}{a(z)c(\zeta) - a(\zeta)c(z)}.$$

A short computation reveals that

$$\kappa_{\mathcal{F}} = 2i\,\mathcal{W}(a, c) \tag{6.3}$$

where $\mathcal{W}(a, c)$ denotes the Wronskian $a\,dc - c\,da$; it follows immediately that $\kappa_{\mathcal{F}}$ is holomorphic. □

6.2. Schwarzians. Let L be a holomorphic line bundle over a Riemann surface X and let Λ be a non-vanishing section of $T^{*(1,0)}(X) \otimes \overline{L}$. With respect to a local coordinate z and a corresponding local non-vanishing holomorphic section η of L we may write $\Lambda = \lambda(z)\, dz \otimes \overline{\eta}$. Replacing z and η by $\widetilde{z} = \phi(z)$ and $\widetilde{\eta} = f\eta$ we find that

$$\lambda(z) = \widetilde{\lambda}(\widetilde{z})\phi'(z)\overline{f}$$

and

$$\begin{aligned}(\log\lambda(z))_{zz} &- \frac{1}{2}(\log\lambda(z))_z^2 \\ &= \left((\log\widetilde{\lambda}(\widetilde{z}))_{\widetilde{z}\widetilde{z}} - \frac{1}{2}(\log\widetilde{\lambda}(\widetilde{z}))_{\widetilde{z}}^2\right)\cdot(\phi'(z))^2 + (\log\phi'(z))_{zz} - \frac{1}{2}(\log\phi'(z))_z^2.\end{aligned}$$

Thus $\mathcal{R}\Lambda \stackrel{\text{def}}{=} (\log\lambda(z))_{zz} - \frac{1}{2}(\log\lambda(z))_z^2$ defines a section of the affine line bundle $A_{\mathcal{S}}(X)$ with transition functions

$$(z,w)\mapsto\left(\phi(z),\ \frac{w-\left((\log\phi'(z))_{zz}-\frac{1}{2}(\log\phi'(z))_z^2\right)}{(\phi'(z))^2}\right).$$

If ω is the area form of a conformal metric then $\mathcal{R}\omega$ is holomorphic if and only if that metric has constant curvature. (See Lemma 15 in §9.)

If f is meromorphic and non-constant then the Schwarzian derivative $\mathcal{S}f \stackrel{\text{def}}{=} \mathcal{R}(df)$ defines a meromorphic section of $A_{\mathcal{S}}(X)$. The standard transformation law **[Leh**, II.1.1] may be written

$$\mathcal{S}(T\circ f) - \mathcal{S}f = (\mathcal{S}T\circ f)\,(df)^2;$$

here T is a non-constant meromorphic function on a domain containing the range of f, the subtraction of two sections of $A_{\mathcal{S}}(X)$ on the left-hand side results in a section of the associated line bundle $\left(T^{*(1,0)}(X)\right)^2$ of quadratic differentials, and $\mathcal{S}T$ is the "classical" (scalar-valued) Schwarzian derivative of T. Note that $\mathcal{S}(T\circ f)\equiv\mathcal{S}f$ if and only if $\mathcal{S}T\equiv 0$ if and only if T is a linear fractional transformation.

A result of Laine and Sorvali [**LaSo**, Cor. 4.8] states that if X is simply-connected then a meromorphic section τ of $A_{\mathcal{S}}(X)$ is the Schwarzian derivative of a non-constant meromorphic function on X if and only if the following condition holds:

(LS) at each pole ζ of τ there is a holomorphic coordinate z vanishing at ζ and an integer $k>1$ so that the representation of τ with respect to z takes the form

$$\frac{1-k^2}{2}z^{-2}+c_{-1}z^{-1}+c_0+\cdots+c_{k-2}z^{k-2}+\cdots \tag{6.4}$$

with

$$\det\begin{pmatrix} 2(1-k) & 0 & 0 & \cdots & 0 & c_{-1} \\ c_{-1} & 2(4-2k) & 0 & \cdots & 0 & c_0 \\ c_0 & c_{-1} & 2(9-3k) & \cdots & 0 & c_1 \\ c_1 & c_0 & c_{-1} & \cdots & 0 & c_2 \\ \vdots & \vdots & \vdots & \ddots & \vdots & \vdots \\ c_{k-3} & c_{k-4} & c_{k-5} & \cdots & 2((k-1)^2-(k-1)k & c_{k-3} \\ c_{k-2} & c_{k-3} & c_{k-4} & \cdots & c_{-1} & c_{k-2} \end{pmatrix} = 0.$$

If the condition in (LS) holds at ζ for a fixed coordinate z then it will hold (with the same value of k) for any other holomorphic coordinate vanishing at ζ; k is in fact the multiplicity at ζ of any solution f of $\mathcal{S}f = \tau$.

Returning now to the notation of the proof of Proposition 8 let us examine the function $C_z = a(z)/c(z)$. It is easy to check that replacing w by $M(z)w + B(z)$ in (6.1) induces no change in $a(z)/c(z)$, whereas changing the representation (6.1) by replacing C by $T(C)$ for some fixed linear fractional transformation T has the effect of replacing $a(z)/c(z)$ by $T\,(a(z)/c(z))$. Thus a/c is determined up to post-composition with a linear fractional transformation by the foliation $\mathcal{F}$.

In view of (6.1) and (6.3), a/c is constant if and only if $\kappa_{\mathcal{F}} \equiv 0$ if and only if the infinity-section $w \equiv \infty$ is a leaf of $\mathcal{F}$. If this does not occur then the Schwarzian derivative $\mathcal{S}(a/c)$ gives rise to a global meromorphic section $\mathcal{S}_{\mathcal{F}}$ of $A_{\mathcal{S}}(X)$ satisfying the condition (LS).

Lemma 9. *Let ν_1, ν_2, ν_3 be distinct local sections of A with graphs contained in leaves of $\mathcal{F}$. Then $\mathcal{S}_{\mathcal{F}} = \mathcal{S}\left(\frac{\nu_2-\nu_3}{\nu_1-\nu_3}\right)$.*

Proof. We may assume that ν_1, ν_2, ν_3 are defined by (6.2) with $C = 0, \infty, 1$, respectively. Then $\frac{\nu_2-\nu_3}{\nu_1-\nu_3} = \frac{a}{c}$. □

Lemma 10. *If $\kappa_{\mathcal{F}}$ is not $\equiv 0$ then the only overshear from A to A taking $\mathcal{F}$ to $\mathcal{F}$ is the identity map.*

Proof. Choose ζ_1, ζ_2 so that ν_{ζ_1} and ν_{ζ_2} are distinct meromorphic sections of A defined on a open set U containing ζ_1 and ζ_2.

Then for generic $\zeta \in U$ the overshear in question must fix the two distinct finite points $\nu_{\zeta_1}(\zeta)$ and $\nu_{\zeta_2}(\zeta)$, forcing the overshear to be the identity map. □

Proposition 11. *If X is a Riemann surface then the map $[\mathcal{F}] \mapsto \mathcal{S}_{\mathcal{F}}$ is a bijection from*

$$\frac{\bigcup_{A \text{ affine over } X} \{\textit{holomorphic foliations } \mathcal{F} \textit{ of } \widehat{A} \textit{ transverse to fibers with } \kappa_{\mathcal{F}} \not\equiv 0\}}{\{\textit{overshears}\}}$$

to

$$\{\textit{meromorphic sections of } A_{\mathcal{S}}(X) \textit{ satisfying } (\mathrm{LS})\}.$$

Proof. The preceding discussion shows that our map is well-defined.

To prove injectivity, note that if $\mathcal{F}$ and $\widetilde{\mathcal{F}}$ are two candidate foliations with $\mathcal{S}_{\widetilde{\mathcal{F}}} = \mathcal{S}_{\mathcal{F}}$ then for a pair of representations of the form (6.1) on the same coordinate patch we have $\mathcal{S}(\widetilde{a}/\widetilde{c}) = \mathcal{S}(a/c)$, so that $\widetilde{a}/\widetilde{c} = T \circ (a/c)$ for some linear fractional transformation T; changing the representation of $\widetilde{\mathcal{F}}$ by replacing $\widetilde{C}$ by $T\widetilde{C}$ we may arrange that $\widetilde{a}/\widetilde{c} = a/c$. Then an elementary calculation shows that the overshear $(z, w) \mapsto \left(z, \frac{a\widetilde{d}-c\widetilde{b}}{d\widetilde{a}-b\widetilde{c}}w + \frac{b\widetilde{d}-d\widetilde{b}}{d\widetilde{a}-b\widetilde{c}}\right)$ maps $\mathcal{F}$ to $\widetilde{\mathcal{F}}$. By Lemma 10 there are no nontrivial locally defined overshears mapping $\mathcal{F}$ to $\mathcal{F}$, so the overshears mapping $\mathcal{F}$ to $\widetilde{\mathcal{F}}$ are locally unique and therefore patch together to define a global overshear.

To prove surjectivity, recall from §6.2 that given any meromorphic section τ of $A_{\mathcal{S}}(X)$ satisfying (LS) and any $\zeta \in X$ we may pick f holomorphic and non-constant on a neighborhood V of ζ with $\mathcal{S}f = \mathcal{S}f^{-1} = \tau$ on V. Then the foliation $\mathcal{F}_f$ on $V \times \mathbb{C}$ defined by $\frac{w}{f(z)w+1} = C$ satisfies $\mathcal{S}_{\mathcal{F}} = \tau$. The argument of the preceding paragraph shows that $\mathcal{F}_f$ is determined by τ up to overshears, so choosing a family of local solutions covering X the overshears defined on overlaps can be used to construct the desired affine bundle A and foliation $\mathcal{F}$. □

Proposition 12. *For a fixed line bundle L over a Riemann surface X and a fixed meromorphic section τ of $A_{\mathcal{S}}(X)$, the map $[\mathcal{F}] \mapsto \kappa_{\mathcal{F}}$ is a bijection from*

$$\frac{\displaystyle\bigcup_{A \text{ associated to } L} \{\text{holomorphic foliations } \mathcal{F} \text{ of } \widehat{A} \text{ tranvserse to fibers with } \mathcal{S}_{\mathcal{F}} = \tau\}}{\{L\text{-shears}\}}$$

to

$$\big\{f \in H\big(T^{*(1,0)(X)} \otimes L^{-1}\big) : f \text{ vanishes to order } k-1 \text{ at } \zeta \Leftrightarrow (6.4) \text{ holds at } \zeta\big\}.$$

Proof. The map in question is well-defined since L-shears do not affect the construction of $\operatorname{Res}_\zeta \nu_\zeta$.

To see that our map takes its values in the prescribed space, note that the quotient rule $d(a/c) = \frac{\mathcal{W}(a,c)}{c^2}$ in conjunction with (6.3) shows that $\kappa_{\mathcal{F}}$ must vanish to order $k-1$ when a/c has multiplicity k, and (as mentioned earlier) this in turn will happen precisely when (6.4) holds at the point in question.

Using the notation of §2.1 it is easy to see that any overshear between affine bundles A_1 and A_2 associated to L is equivalent modulo L-shears to a map $F_g : (L, J_\omega) \to (L, J_{g\omega})$ that dilates each fiber L_ζ by the factor $g(\zeta)$; here g is a holomorphic map from X into $\mathbb{C} \setminus \{0\}$. The bijectivity claimed in Proposition 12 now follows easily from Proposition 11 and the transformation law $\kappa_{F_g^*\mathcal{F}} = g^{-1}\kappa_{\mathcal{F}}$. □

6.3. Recognizing extended Levi-foliations. In §9 we will prove the following.

Addendum to Theorem 7. *If $\mathcal{F}$ is the extended Levi-foliation of a Levi-flat hypersurface $S_{\gamma,h}$ then $\kappa_{\mathcal{F}} = -2ih^2\overline{\omega}$ and $\mathcal{S}_{\mathcal{F}} = \mathcal{R}(2ih^2\omega \wedge \overline{\omega}) = \mathcal{R}(\overline{\omega})$, where $\omega = \bar{\partial}\gamma$.*

Suppose we are given a rank 1 affine bundle A over a Riemann surface X and a holomorphic foliation $\mathcal{F}$ of $\widehat{A}$ transverse to fibers. How can we determine whether or not $\mathcal{F}$ is the extended Levi-foliation $\mathcal{F}_{S_{\gamma,h}}$ for some Levi-flat $S_{\gamma,h}$?

If X is non-compact then Theorem 6, Proposition 11, and the Addendum to Theorem 7 combine to yield the (somewhat tautological) conclusion that $\mathcal{F}$ is an extended Levi-foliation if and only if

- $\kappa_{\mathcal{F}} \equiv 0$ and $\mathcal{F}$ has a leaf with unitary holonomy projecting bijectively onto X,

or

- $\mathcal{S}_{\mathcal{F}} = \mathcal{R}(\Lambda)$ where Λ is the area form of a conformal metric on X of curvature -1 with all total angles $\in 2\pi\mathbb{N}$.

If X is compact then Theorem 5 shows that $\mathcal{F}$ will not be an extended Levi-foliation unless the degree of the corresponding line bundle L is ≥ 0.

If $\deg L = 0$ then the only possible extended Levi-foliation is that on $\widehat{A} = \widehat{L}$ induced via condition (CM) by the unique (up to positive constants) flat metric on L.

If $\deg L > 0$ then Theorem 5, Proposition 12 and the Addendum to Theorem 7 combine to show that $\mathcal{F}$ is an extended Levi-foliation if and only if $\mathcal{S}_{\mathcal{F}} = \mathcal{R}(\Lambda)$, where Λ is the area form of the metric $\widetilde{h}$ constructed from the divisor of κ_ω in the proof of Theorem 5.

7. Foliations from "phantom hypersurfaces"

Let A be an affine bundle associated to the cotangent bundle $T^{*(1,0)}(\mathbb{CP}^1)$ of the Riemann sphere $\mathbb{CP}^1$. Since $\deg T^{*(1,0)}(\mathbb{CP}^1) = -2$, Theorem 5 shows that A does not contain a Levi-flat hypersurface with circular fibers. On the other hand, if ω is the area form for the usual spherical metric on $\mathbb{CP}^1$ then taking $A = (L, J_\omega)$, $\gamma = 0$, and h to be the metric on $T^{*(1,0)}(\mathbb{CP}^1)$ induced by the spherical metric on $\mathbb{CP}^1$, it turns out that the condition (LHM) from Theorem 7 still defines a holomorphic foliation $\mathcal{F}$ on $\widehat{A}$ transverse to fibers. Comparing with the last paragraph of §4 we see that $\mathcal{F}$ is formally $\mathcal{F}_{\Sigma_{\omega,ih}}$ – but of course the radius of the fibers is not allowed to be imaginary!

More generally we have the following.

Theorem 13. *Let A be an affine bundle over X associated to a line bundle L equipped with metric h, and let γ be a smooth section of A. Suppose that $\omega = \bar\partial\gamma$ is nowhere-vanishing. Then the following conditions are equivalent:*

(1) *Condition* (LHM) *of Theorem 7 describes a holomorphic foliation $\mathcal{F}$ of $\widehat{A}$ with leaves transverse to fibers.*

(2) $\kappa_\omega \overset{def}{=} -\frac{i\partial\bar\partial \log\omega}{\omega}$ *is a holomorphic section of $L^{-1} \otimes T^{*(1,0)}(X)$ and $h^2|\omega|$ is a flat metric on $L \otimes T^{(1,0)}(X)$.*

The notation κ_ω is motivated by the fact that if ω is a positive (1,1)-form then κ_ω is the curvature of the conformal metric with ω as area form. The notation $\kappa_{\mathcal{F}}$ from §6.1 was motivated by the Addendum to Theorem 14 found at the end of this section.

Theorem 13 will be proved in §9 essentially as a special case of the following result which allows for zeroes of ω.

Theorem 14. *Let A be an affine bundle over X associated to a line bundle L equipped with metric h, let γ be a smooth section of A, and let $\omega = \bar{\partial}\gamma$. Assume that ω is not $\equiv 0$. Then the following conditions are equivalent:*

(1) *Condition* (LHM) *of Theorem 7 describes a holomorphic foliation $\mathcal{F}$ of $\widehat{A}$ with leaves transverse to fibers.*

(2) *There is an open set $U \subset A$ with $\pi(U) = X$ such that condition* (LHM) *of Theorem 7 describes a holomorphic foliation $\mathcal{F}$ of U with leaves transverse to fibers.*

(3) *For each $\zeta \in X$ there is a neighborhood V of p together with*

 (a) *a holomorphic section ν of A on V with $\nu - \gamma$ non-vanishing*

 and

 (b) *a non-vanishing holomorphic section η of L on V*

 such that $\log \|\nu - \gamma\|$ *and $\eta/(\nu - \gamma)$ are harmonic.* (*Note that $\eta/(\nu - \gamma)$ will in general be $\mathbb{C}$-valued.*)

(4) *ω and h admit local representations of the form*

$$\omega = \frac{\eta\, d\bar{g}}{(f - \bar{g})^2}$$
$$h = |f - \bar{g}|\, \mu;$$

 here f and g are holomorphic functions with $f - \bar{g}$ non-vanishing, η is a non-vanishing holomorphic section of L and μ is a flat metric on L.

(5) (a) *$h^2|\omega|$ is a flat metric on $L \otimes T^{(1,0)}(X)$ off of the zero set of ω;*

 (b) *$\mathcal{R}\overline{\omega}$ is a meromorphic section of $A_{\mathcal{S}}(X)$ satisfying condition* (LS) *from* §6.2;

 (c) *if $\mathcal{R}\overline{\omega}$ has a pole at ζ and z is a local coordinate vanishing at ζ then*

$$\omega = \bar{z}^{k-1}\varphi \tag{7.1}$$

 for some smooth φ defined near ζ with $\phi(\zeta) \neq 0$.

Remark 1. In condition (5) above, the values of k in (7.1) and condition (LS) will coincide wherever ω vanishes.

When $\omega \equiv 0$, the existence of section ν (not $\equiv \gamma$) satisfying the condition in (LHM) implies that h is flat. In this case (CM) defines a foliation but (LHM) does not.

Addendum to Theorem 14. *For $\mathcal{F}$ as in Theorem 14 we have*

$$\kappa_{\mathcal{F}} = \kappa_{\omega},$$
$$\mathcal{S}_{\mathcal{F}} = \mathcal{R}(\partial\bar{\partial}\log\omega)$$

on $\{\zeta \in X : \omega(\zeta) \neq 0\}$.

Theorem 14 is proved in the next section; the Addendum will be proved in §9.

Remark 2. A Levi-flat $S \subset A$ given by the equation $h^2\,|w-\gamma|^2 = 1$ is the pullback via the map $\mathrm{Id}\times_X\overline{\mathrm{Id}} : A \to A\times_X \overline{A}$ of the hypersurface $h^2(w-\gamma)(\widetilde{w}-\overline{\gamma}) = 1$. In view of the Addendum to Theorem 7, a foliation $\mathcal{F}$ constructed from the condition (LHM) may be viewed as stemming from the hypersurface $i\kappa_{\mathcal{F}}(w-\gamma)(\widetilde{w}-\overline{\gamma}) = 2\,\partial\overline{\gamma}$, though this hypersurface may not intersect $\left(\mathrm{Id}\times_X\overline{\mathrm{Id}}\right)(A)$.

8. Proof of Theorem 14

(1) ⇒ (2): Take $U = X$. □

(2) ⇒ (3): According to (2), for $\zeta \in X$ we may find a neighborhood V of ζ and a biholomorphic map Ψ from $V\times\Delta$ to an open subset of A with $\pi\circ\Psi = \pi$ such that for fixed ξ, $\Psi(\cdot,\xi)$ is a holomorphic section with $\log\|\Psi(\cdot,\xi)-\gamma(\cdot)\|$ harmonic. (Here Δ is the unit disk in $\mathbb{C}$.)

Let $\nu(\cdot) = \Psi(\cdot,0)$ and let η be the non-vanishing holomorphic section of L over V given by $z \mapsto \frac{d}{d\xi}\Psi(z,\xi)\big|_{\xi=0}$.

Working with bundle coordinates (z,w) on $\pi^{-1}(V)$ we may write $\Psi(z,\xi) = (z,\psi(z,\xi))$, $h = e^{u(z)}\,|dw|$ to obtain

$$\begin{aligned} 0 &= \partial\bar{\partial}\log\left(e^{2u(z)}\,|\psi(z,\xi)-\gamma(z)|^2\right)\\ &= 2\,\partial\bar{\partial}u + \partial\bar{\partial}\log(\psi-\gamma) + \partial\bar{\partial}\log\overline{(\psi-\gamma)} \end{aligned}$$

for each fixed ξ. It follows that

$$\begin{aligned} 0 &= 2\,\partial\bar{\partial}\left(\frac{d}{d\xi}u\Big|_{\xi=0}\right) + \partial\bar{\partial}\left(\frac{d}{d\xi}\log(\psi-\gamma)\Big|_{\xi=0}\right) + \partial\bar{\partial}\left(\frac{d}{d\xi}\log\overline{(\psi-\gamma)}\Big|_{\xi=0}\right)\\ &= 0 + \partial\bar{\partial}(\eta/(\nu-\gamma)) + 0, \end{aligned}$$

as required. □

(3) ⇒ (4): We may locally represent $\eta/(\nu-\gamma)$ as $f-\overline{g}$ with f, g holomorphic, $f-\overline{g}$ non-vanishing. Thus

$$\nu - \gamma = \frac{\eta}{f-\overline{g}}$$

and

$$\omega = -\bar{\partial}(\nu - \gamma) = \frac{\eta\, d\bar{g}}{(f - \bar{g})^2}.$$

Moreover,

$$v \stackrel{\text{def}}{=} \log \|\nu - \gamma\| = \log h|\nu - \gamma| = \log \frac{h|\eta|}{|f - \bar{g}|}$$

is harmonic, so

$$h = |f - \bar{g}|\mu$$

where $\mu \stackrel{\text{def}}{=} e^{v}|\eta|^{-1}$ is a flat metric on L. □

(4) ⇒ (1): Suppose that in one of the hypothetical local representations the function g is constant. Then $\omega \equiv 0$ on the open set in question, and an analytic continuation argument shows $\omega \equiv 0$ globally, contrary to hypothesis. So g must be non-constant in each of the hypothetical local representations.

Let ν be a holomorphic section of A with $\log \|\nu - \gamma\|$ harmonic on a connected open set $V \subset X$ on which ω and h admit the prescribed representations. Since

$$\bar{\partial} \frac{-\eta}{f - \bar{g}} = \omega = \bar{\partial}\gamma$$

we have that

$$\rho \stackrel{\text{def}}{=} \nu - \gamma - \frac{\eta}{f - \bar{g}}$$

is a holomorphic section of L.

Thus

$$\begin{aligned} \|\nu - \gamma\| &= h \cdot |\nu - \gamma| \\ &= |f - \bar{g}|\mu \cdot \left|\rho + \frac{\eta}{f - \bar{g}|}\right| \\ &= \mu\, |(f - \bar{g})\rho + \eta|\,. \end{aligned}$$

On the set

$$V' = \{\zeta \in V : \rho(\zeta) \neq 0\}$$

we have

$$\|\nu - \gamma\| = \mu|\rho|\,|(f - \bar{g}) + \eta/\rho|\,.$$

Since g is non-constant, the set

$$V'' \stackrel{\text{def}}{=} \{\zeta \in V' : \frac{\eta}{\rho}(\zeta) + f(\zeta) \neq \overline{g(\zeta)}\}$$

is dense in V'. On V'' we have

$$\begin{aligned}
0 &= \partial\bar{\partial}\log\|\nu-\gamma\| \\
&= \frac{1}{2}\partial\bar{\partial}\log\left(\frac{\eta}{\rho}+f-\bar{g}\right)+\frac{1}{2}\partial\bar{\partial}\log\overline{\left(\frac{\eta}{\rho}+f-\bar{g}\right)} \quad \text{(since } \bar{\partial}\rho=0 \text{ and } \mu \text{ is flat)} \\
&= \frac{d\left(\frac{\eta}{\rho}+f\right)\wedge\overline{dg}}{2\left(\frac{\eta}{\rho}+f-\bar{g}\right)^2}+\frac{dg\wedge\overline{d\left(\frac{\eta}{\rho}+f\right)}}{2\left(\overline{\frac{\eta}{\rho}+f}-g\right)^2}
\end{aligned}$$

so

$$\frac{d\left(\frac{\eta}{\rho}+f\right)\wedge\overline{dg}}{\left(\frac{\eta}{\rho}+f-\bar{g}\right)^2}=-\frac{dg\wedge\overline{d\left(\frac{\eta}{\rho}+f\right)}}{\left(\overline{\frac{\eta}{\rho}+f}-g\right)^2}. \tag{8.1}$$

If $\frac{\eta}{\rho}+f$ is non-constant then we may apply $\partial\bar{\partial}\log$ to both sides of (8.1) to obtain

$$\frac{d\left(\frac{\eta}{\rho}+f\right)\wedge\overline{dg}}{\left(\frac{\eta}{\rho}+f-\bar{g}\right)^2}=\frac{dg\wedge\overline{d\left(\frac{\eta}{\rho}+f\right)}}{\left(\overline{\frac{\eta}{\rho}+f}-g\right)^2} \tag{8.2}$$

on a dense subset V''' of V''.

Combining (8.1) and (8.2) we find that

$$\frac{d\left(\frac{\eta}{\rho}+f\right)\wedge\overline{dg}}{\left(\frac{\eta}{\rho}+f-\bar{g}\right)^2}\equiv 0$$

on V''', hence on V''. Since g is non-constant, $\frac{\eta}{\rho}+f$ must be constant on V'', hence also on V'. Thus $\rho=\frac{\eta}{C-f}$ on V'; if ρ does not vanish identically then $\rho=\frac{\eta}{C-f}$ also on V (so in fact $V'=V$). Thus

$$\nu=\gamma+\frac{\eta}{f-\bar{g}}+\frac{\eta}{C-f}; \tag{8.3}$$

taking $C=\infty$ we obtain the remaining case $\rho\equiv 0$, $\nu=\gamma+\frac{\eta}{f-\bar{g}}$.

This describes the required foliation on $\widehat{A}\big|_V$, and the local uniqueness shows that these foliations patch together to give the required foliation on $\widehat{A}$. □

(4) ⇒ (5): $h^2|\omega|=\mu^2|\eta||d\bar{g}|$ is flat off of the zero set of dg. Straightforward computation (see (9.1) below) shows that $\mathcal{R}\bar{\omega}=\mathcal{R}(dg)=\mathcal{S}g$ so that **[LaSo**, Cor. 4.8**]** shows that $\mathcal{R}\bar{\omega}$ satisfies (LS). Condition (5c) holds by inspection. □

(5) ⇒ (4): Let $\zeta \in X$. By **[LaSo**, Cor. 4.8] there is a meromorphic function g defined near ζ with

$$\mathcal{R}\overline{\omega} = \mathcal{R}(dg) = \mathcal{S}g; \tag{8.4}$$

if $\mathcal{R}\overline{\omega}$ has a pole at ζ then the multiplicity of g at ζ is the integer k from (6.2). Post-composing g with a fractional linear transformation we may assume that $g(\zeta) = 0$.

We focus first on the case where $\mathcal{R}\overline{\omega}$ has no pole at ζ. Then we may take g to be our local coordinate z. Fixing a local non-vanishing holomorphic section ξ of L we set $\omega = \lambda(z)\, d\overline{z} \otimes \xi$.

We wish to arrange that $(\log \overline{\lambda})_z(\zeta) \neq 0$. If this is not true we may replace g by $\frac{g}{1+g}$. We find then that $\lambda(z)$ is replaced by $(1+\overline{z})^{-2}\lambda(z)$ and that $(\log \overline{\lambda})_z$ is replaced by $-2(1+z) + (1+z)^2(\log \overline{\lambda})_z$ so that $(\log \overline{\lambda})_z(\zeta)$ is now non-zero.

With our choice of coordinate now fixed we have

$$\mathcal{R}\overline{\omega} = \mathcal{S}g = \mathcal{S}z \equiv 0$$

and so

$$\left(\frac{1}{(\log \overline{\lambda})_z}\right)_z = -\frac{(\log \overline{\lambda})_{zz}}{(\log \overline{\lambda})_z^2} = -\frac{1}{2}.$$

Thus

$$\frac{1}{(\log \overline{\lambda})_z} = -\frac{z - \overline{f}}{2}$$

with f holomorphic, $f(p) \neq \overline{g}(p)$. This yields

$$(\log \overline{\lambda})_z = -\frac{2}{z - \overline{f}}$$

and hence

$$\log \overline{\lambda} = -2\log(z - \overline{f}) + \overline{h}, \tag{8.5}$$

h holomorphic. Then

$$\omega = \frac{e^h \xi\, \overline{dg}}{(f - \overline{g})^2};$$

setting $\eta = e^h \xi$ we have the desired local representation for ω, and condition (5a) sets up the corresponding representation for h.

We turn now to the case where $\mathcal{R}\overline{\omega}$ has a pole at ζ, recalling that in this case g has a zero of multiplicity k at ζ, where k is the integer from (6.2).

Here g cannot serve as a coordinate at ζ but using g as a (non-univalent) coordinate z in a punctured neighborhood of ζ and replacing z by $\sqrt[k]{z}$ in (5c) to obtain

$$\lambda(z) = k^{-1}\varphi(z^{1/k}),\ \varphi(0) \neq 0 \tag{8.6}$$

we have

$$(\log \lambda)_{\overline{z}} = \mathcal{O}\left(z^{\frac{1}{k}-1}\right). \tag{8.7}$$

If $(\log \lambda)_{\overline{z}} \to 0$ as we approach ζ then replacing g by $\frac{g}{1+g}$ as before we may arrange that $(\log \lambda)_{\overline{z}}$ tends to a non-zero limit as we approach ζ; thus we may assume that

$$(\log \lambda)_{\overline{z}} \text{ does not approach } 0 \text{ at } \zeta. \tag{8.8}$$

Following our previous computation we set

$$f(z) = \begin{cases} -\overline{z} - \frac{2}{(\log \lambda)_{\overline{z}}} & \text{for } (\log \lambda)_{\overline{z}} \neq 0 \\ \infty & \text{for } (\log \lambda)_{\overline{z}} = 0. \end{cases}$$

so that

$$\frac{1}{f(z)} = \begin{cases} -\frac{(\log \lambda)_{\overline{z}}}{2+\overline{z}(\log \lambda)_{\overline{z}}} & \text{for } (\log \lambda)_{\overline{z}} \neq 0 \\ 0 & \text{for } (\log \lambda)_{\overline{z}} = 0. \end{cases}$$

Our earlier work shows that the continuous function $1/f$ is holomorphic where it is non-zero, so Radó's Theorem [**Nar**, 11.8] shows that $1/f$ is in fact holomorphic in a deleted neighborhood of p. In view of (8.7) and (8.8), after shrinking our neighborhood we may assume that f is holomorphic in a neighborhood of ζ with $f \sim cg^{j/k}$ for some $c \neq 0$ and some integer $0 \leq j \leq k-1$. Consequently we may also assume that $f \neq \overline{g}$ except perhaps at ζ.

Continuing on to (8.5) we find that we must replace h by $h + \frac{2j}{k}\log g$ to ensure that h is single-valued near ζ. We thus have

$$\omega = \frac{g^{2j/k} e^h \xi \, \overline{dg}}{(f - \overline{g})^2}$$

with h holomorphic in a punctured neighborhood of ζ. In view of (8.6), h must have a removable singularity at ζ. Letting $\eta = e^h \xi$ as before we find that

$$\omega = \frac{g^{2j/k} \eta \, \overline{dg}}{(f - \overline{g})^2}.$$

The smoothness of φ in condition (5c) implies that of

$$\frac{f}{g^{j/k}} - \sqrt{\frac{\eta \, \overline{dg}}{\omega}} = \frac{\overline{g}}{g^{j/k}};$$

differentiating k times it follows that $j = 0$; thus $f \neq \overline{g}$ at ζ.

As before, the local representation for ω and condition (5a) induces the corresponding representation for h. □

9. More proofs

Proof of Addendum to Theorem 14. Comparing the formula (8.3) describing $\mathcal{F}$ to the definitions in §§6.1 and 6.2 we find that

$$\begin{aligned} \nu_\zeta &= \gamma + \frac{\eta}{f-\overline{g}} + \frac{\eta}{f(\zeta)-f} \\ \operatorname*{Res}_\zeta \nu_\zeta &= -\frac{\eta}{df} \\ \kappa_{\mathcal{F}} &= \frac{2i\,df}{\eta} \\ \mathcal{S}_{\mathcal{F}} &= \mathcal{S} f. \end{aligned}$$

On the other hand, from the representation in condition (4) of Theorem 14 we have

$$\begin{aligned} \partial\bar\partial \log\omega &= -2\frac{df\wedge d\overline{g}}{(f-\overline{g})^2} \\ \kappa_\omega &= 2i\frac{df}{\eta} \\ \mathcal{R}(\partial\bar\partial\log\omega) &= \mathcal{S} f. \end{aligned}$$

(The computation of $\mathcal{R}(\partial\bar\partial\log\omega)$ is facilitated by taking f to be the coordinate function z – permissible away from critical points of f – leading to

$$\mathcal{R}(\partial\bar\partial\log\omega) = \frac{2}{(f-\overline{g})^2} - \frac{1}{2}\left(\frac{-2}{f-\overline{g}}\right)^2 = 0 = \mathcal{S} f \tag{9.1}$$

as claimed.)

Thus everything matches. □

Lemma 15. *If ω is a non-vanishing section of $L\otimes T^{*(0,1)}(X)$ then $\mathcal{R}\overline{\omega}$ is holomorphic if and only if κ_ω is holomorphic.*

Proof. This follows from the easily-checked identity

$$\left(\mathcal{R}\overline{\omega}\right)_z = i\omega\,(\kappa_\omega)_{\overline{z}}. \tag{9.2}$$

□

Proof of Theorem 13. This is now immediate from conditions (1) and (5) from Theorem 14 combined with Lemma 15. □

Proof of Theorem 7 *and Addendum to Theorem* 7. If h is flat then in local coordinates we have $h = e^{\operatorname{Re} f(z)}\,|dw|$, f holomorphic. The hypersurface is given by $e^{\operatorname{Re} f(z)}|w - \gamma(z)| = 1$, the leaves of the Levi-foliation are defined by equations of the form $e^{f(z)}(w-\gamma(z)) = e^{i\theta_0}$, and leaves of the extended foliation are given by equations of the form $e^{f(z)}(w-\gamma(z)) = C$. If ν is holomorphic then the holomorphic function

$e^{f(z)}(\nu(z)-\gamma(z))$ will be constant if and only if $\|\nu-\gamma\| = \left|e^{f(z)}(\nu(z)-\gamma(z))\right|$ is constant.

If h is not flat then $h^2\overline{\omega}$ is holomorphic so that condition (5a) of Theorem 14 holds. Also, away from zeros of ω we have $0 = \partial\bar{\partial}\log h^2\omega$ so that

$$\begin{aligned}\partial\bar{\partial}\log\omega &= -2\partial\bar{\partial}\log h = \Theta \\ &= 2h^2\omega\wedge\overline{\omega}\end{aligned}$$

and $\kappa_\omega = -2ih^2\overline{\omega}$ is holomorphic. By Lemma 15, $\mathcal{R}\overline{\omega}$ is holomorphic away from zeroes of ω.

The fact that $h^2\overline{\omega}$ is holomorphic also easily implies that condition (5c) holds and that $\mathcal{R}\overline{\omega}$ has poles at zeroes of ω.

Consulting Theorem 13 we see that the condition (LHM) provides the desired extended foliation away from zeroes of ω. By the Addendum to Theorem 14 we have $\kappa_{\mathcal{F}} = \kappa_\omega = -2ih^2\overline{\omega}$ and $\mathcal{S}_{\mathcal{F}} = \mathcal{R}\left(\partial\bar{\partial}\log\omega\right) = \mathcal{R}\left(2h^2\omega\wedge\overline{\omega}\right) = \mathcal{R}(\overline{\omega})$.

If $\omega(z_0) = 0$ then picking distinct sections ν_1, ν_2, ν_3 near z_0 whose graphs are leaves of the original Levi-foliation we have from Lemma 9 that $\mathcal{R}(\overline{\omega}) = \mathcal{S}_{\mathcal{F}} = \mathcal{S}\left(\frac{\nu_2-\nu_3}{\nu_1-\nu_3}\right)$ so that $\mathcal{R}(\overline{\omega})$ satisfies condition (LS) from §6.2. Thus condition (5) of Theorem 14 holds, so condition (1) must hold as well. □

References

[AnLe] E. Andersén and L. Lempert, On the group of holomorphic automorphisms of $\mathbb{C}^n$, Invent. Math. 110 (1992), 371–388.

[Bar] D. Barrett, On a class of conformal metrics arising in work of Seiberg and Witten, Math. Z. 233 (2000), 149–164.

[Ber] B. Berndtsson, Levi-flat surfaces with circular sections, Several Complex Variables, Math. Notes 38, Princeton University Press, 1993, 136–159.

[CaLN] C. Camacho and A. Lins Neto, Geometric theory of foliations, Birkhäuser, 1985.

[For] O. Forster, Lectures on Riemann surfaces, Grad. Texts in Math. 81, Springer-Verlag, 1981.

[GrHa] P. Griffiths and J. Harris, Principles of algebraic geometry, John Wiley & Sons, 1978.

[HuTr] D. Hulin and M. Troyanov, Prescribing curvature on open surfaces, Math. Ann. 293 (1992), 277–315.

[Kra] S. Krantz, Function theory of several complex variables (2nd ed.), Wadsworth & Brooks/Cole, 1992.

[LaSo] I. Laine and T. Sorvali, Local solutions of $w'' + A(z)w = 0$ and branched polymorphic functions, Results Math. 10 (1986), 107–129.

[Leh] O. Lehto, Univalent functions and Teichmüller spaces, Grad. Texts in Math. 109, Springer-Verlag, 1987.

[Nar] R. Narasimhan, Complex analysis in one variable, Birkhäuser, 1985.

[Roc] R. Rochberg, Interpolation of Banach spaces and negatively curved vector bundles, Pacific J. Math. 110 (1984), 355–376.

[SuTh] D. Sullivan and W. Thurston, Extending holomorphic motions, Acta Math. 157 (1986), 243–257.

Department of Mathematics
University of Michigan
Ann Arbor, MI 48109-1109, U.S.A.
barrett@umich.edu

Weighted estimates for the $\bar{\partial}$-equation

Bo Berndtsson

1. Introduction

In this paper we will survey a number of variations on the theme of Hörmander type weighted L^2-estimates for the $\bar{\partial}$-equation. The original Hörmander theorem, **[H]**, gives an estimate for the weighted L^2-norm

$$\int |u|^2 e^{-\phi}$$

of a solution, u, to the equation $\bar{\partial} u = f$, where ϕ is a plurisubharmonic weight function. The possibility to change weight makes this theorem a very flexible tool in the construction of holomorphic functions with various prescribed properties. In this paper we shall consider similar estimates containing several weight functions, sometimes not plurisubharmonic. These results have been obtained by a number of authors, **[DF]**, **[Oh]**, **[DH]**, **[DOh]**, **[S1]**, **[McN1]**, **[McN3]**, **[B1]**, **[B2]**, starting with the groundbreaking work of Donnelly and Fefferman. They have several applications, the most noteworthy being perhaps the extension theorem of Ohsawa and Takegoshi **[OhT]**.

In our discussion of these results we shall pay particular attention to when it is possible to find one fixed solution that satisfies weighted estimates with respect to a class of different weight functions, **[B1]**, **[B2]**, **[Del]**, **[BCh]**. This permits even finer control of the solution, and we shall see that in some cases it leads to sup-norm estimates. Moreover, the fixed solution that we find will be the solution of minimal norm in one particular L^2-space, and the results are therefore of use in connection with the Bergman kernel, **[Del]**, **[BCh]**.

I would like to stress that this is principally a survey of known material. The new results are Theorems 2.8, 4.2 and 4.5.

2. L^2-estimates

Throughout this paper we will let D be a bounded pseudoconvex domain in $\mathbb{C}^n$, and we shall study the equation

$$\bar{\partial} u = f \tag{2.1}$$

Complex Analysis and Geometry
Ohio State Univ. Math. Res. Inst. Publ. 9

where f is a $\bar\partial$-closed (0, 1)-form in D. Most of what we have to say extends to forms of other bidegrees, forms with values in complex line bundles, and to domains in complex manifolds. Sometimes we shall indicate what modifications are needed in these more general situations, but principally we stick to the simplest case. In the proofs we give we shall also assume that all the forms f and all the weight functions extend smoothly to the boundary of D, which we also assume to be smooth. The final estimates that we obtain will however not depend on these smoothness assumptions.

Let $\Theta = i\Sigma\Theta_{j,k}d\zeta_j \wedge d\bar\zeta_k$ be a continuous strictly positive (1, 1)-form in D, i.e., assume that the matrix $(\Theta_{j,k})$ is strictly positive in every point. The pointwise norm of a (0, 1)-form, f measured in the metric defined by Θ is then defined by

$$|f|^2_\Theta = \Sigma\Theta^{j,k} f_j \bar f_k,$$

where $(\Theta^{j,k})$ is the matrix inverse to $(\Theta_{j,k})$.

Let ϕ be a plurisubharmonic function in D. The original Hörmander estimate then says that there is a solution to (2.1) such that

$$\int |u|^2 e^{-\phi} \le \int |f|^2_{\partial\bar\partial\phi}\, e^{-\phi}. \tag{2.2}$$

(This estimate is only implicit in **[H]**, where the norm $|f|^2_\Theta$ is replaced by $|f|^2/c$, with c the smallest eigenvalue of Θ. The precise formulation (2.2) can be found in **[Dem]**).

The inequality (2.2) means in particular that the more plurisubharmonic ϕ is, i.e., the bigger $\partial\bar\partial\phi$ is, the more we gain in the estimate. If however we want to chose e.g. $\phi = 0$, (2.2) is in principle void. If D is bounded this is of course no problem, since we can introduce an artificial weight function $\phi = |z|^2$, that will change the final estimate only by a constant if the domain is bounded.

The important discovery in **[DF]** is that one can do better. It is not necessary that ϕ is bounded, it suffices that

$$i\partial\phi \wedge \bar\partial\phi \le ri\partial\bar\partial\phi, \tag{2.3}$$

where r is some constant. The following theorem is essentially contained in **[DF]**.

Theorem 2.1. *Let ϕ and ψ be plurisubharmonic functions of class C^2 in D, and assume that ϕ satisfies condition* (2.3). *Then there is a solution, u, to* (2.1) *satisfying the estimate*

$$\int |u|^2 e^{-\psi} \le Cr \int |f|^2_{\partial\bar\partial\phi} e^{-\psi},$$

where C is an absolute constant.

One case of interest and the original case in **[DF]** is when $D = B$ is the unit ball in $\mathbb{C}^n$ and $\phi = \log 1/(1 - |z|^2)$ (or, more generally, when D is strictly pseudoconvex and $\phi = 1/(n+1)\log B(z, z)$ where B is the Bergman kernel of D). Then ϕ satisfies (2.3) with $r = 1$ and we obtain an estimate of u in terms of the norm of f with respect to the Bergman metric. Note that this estimate would not be possible to obtain by choosing ϕ bounded since there is no bounded function satisfying

$$i\partial\bar{\partial}\phi \geq i\partial\bar{\partial}\log(1/1-|z|^2).$$

We shall now discuss generalizations of Theorem 2.1. We start from the following lemma.

Lemma 2.2. *Let Θ be a positive form, and let ϕ be a plurisubharmonic function satisfying*

$$i\partial\phi \wedge \bar{\partial}\phi \leq r\Theta$$

with $r < 1$. Let ψ be an arbitrary function in, say L^∞. Suppose that for any $\bar{\partial}$-closed form f we can find a solution u to (2.1) *such that*

$$\int |u|^2 e^{-\phi-\psi} \leq \int |f|^2_\Theta e^{-\phi-\psi}. \tag{2.4}$$

Then u_0, the solution to (2.1) *of minimal norm in $L^2(e^{-\psi})$, satisfies*

$$\int |u_0|^2 e^{\phi-\psi} \leq 6/(1-r)^2 \int |f|^2_\Theta e^{\phi-\psi}. \tag{2.5}$$

Proof. Put $u = u_0 e^\phi$. Since

$$\int u_0\bar{h}e^{-\psi} = 0$$

for any square integrable holomorphic function h in D, clearly

$$\int u\bar{h}e^{-\psi}e^{-\phi} = 0$$

for any such h. Thus u is the minimal solution in $L^2(e^{-\phi-\psi})$ to the equation

$$\bar{\partial}u = \bar{\partial}(u_0 e^\phi) = f e^\phi + u_0\bar{\partial}\phi e^\phi.$$

Therefore u must satisfy the estimate in (2.4) viz

$$\int |u|^2 e^{-\phi-\psi} \leq \int |\bar{\partial}(u_0 e^\phi)|^2_\Theta e^{-\phi-\psi},$$

or in other words

$$\int |u_0|^2 e^{\phi-\psi} \leq \int |f + u_0\bar{\partial}\phi|^2_\Theta\, e^{\phi-\psi}.$$

Now we use, for $\tau > 0$,

$$|f + u_0\bar{\partial}\phi|^2_\Theta \leq (1+\tau)|f|^2_\Theta + (1+1/\tau)|u_0|^2|\bar{\partial}\phi|^2_\Theta,$$

and note that the hypothesis on ϕ means precisely that $|\bar{\partial}\phi|^2_\Theta \leq r$. Take $\tau = 2/(1-r)$ and absorb the last member of the right hand side in the left hand side. The result is (2.5). □

One application of Lemma 2.2 is when ϕ and ψ are plurisubharmonic and $\Theta = i\partial\bar\partial\phi$. It then follows from Hörmander's theorem that (2.4) holds. Consequently we get with the same notation as before

Theorem 2.3. *Let ϕ and ψ be plurisubharmonic functions in D and suppose ϕ satisfies* (2.3) *with $r < 1$. Then u_0, the $L^2(e^{-\psi})$-minimal solution to* (2.1), *satisfies*

$$\int |u_0|^2 e^{\phi-\psi} \leq 6/(1-r)^2 \int |f|^2_{\partial\bar\partial\phi} e^{\phi-\psi}. \tag{2.6}$$

Theorem 2.1 is clearly a consequence of (2.6) since we may replace ψ by $\psi+\phi$. After that the estimate becomes homogeneous in ϕ so we can permit r to be any finite number. For (2.6) to hold the assumption $r < 1$ is however necessary - the example showing this is again D equal to the unit ball and $\phi = \log 1/(1-|z|^2)$.

We now turn to the question for which plurisubharmonic functions ϕ and which Θ an estimate of the type

$$\int |u_0|^2 e^{-\phi-\psi} \leq \int |f|^2_\Theta e^{-\phi-\psi}$$

holds for the $L^2(e^{-\psi})$-minimal solution to (2.1). That an estimate of this type holds for *some* solution u is of course a direct consequence of (2.2), but the problem becomes non-trivial if we insist on having a solution independent of ϕ. Here is one result in this direction, **[B1]**. We still denote by u_0 the minimal solution to (2.1) in $L^2(e^{-\psi})$.

Theorem 2.4. *Let ϕ and ψ be plurisubharmonic and suppose ϕ satisfies condition* (2.1) *with $r = 1$. Then*

$$\int |u_0|^2 e^{-\phi-\psi} \leq \int |f|^2_{\partial\bar\partial\psi} e^{-\phi-\psi}.$$

Notice that the assumption on ϕ just means that $w = e^{-\phi}$ is plurisuperharmonic. Even though Theorem 2.4 may seem less surprising than Theorem 2.3 it seems that it is not a formal consequence of the Hörmander theorem in the same way as Theorem 2.3 is. The proof is instead based on an improvement of the basic a priori estimate used in the proof of Hörmander's theorem. There are several alternative formulations of this, see **[DF]**, **[Oh]**, **[S2]**, **[McN1]** and **[B1, B2]**. Here we will use the differential identity from **[B1, B2]**, which is actually a reformulation of a computation from **[S2]**, and there baptized the "$\partial\bar\partial$-Bochner–Kodaira technique".

Let $\alpha = \Sigma\alpha_j d\bar z_j$ be a smooth $(0,1)$-form and let, for a given weight function ϕ, $\bar\partial^*_\phi$ denote the formal adjoint of the $\bar\partial$-operator in $L^2(e^{-\phi})$. Explicitly

$$\bar\partial^*_\phi = -\Sigma e^\phi \frac{\partial}{\partial z_j} e^{-\phi}\alpha_j.$$

We then have (for a proof see **[B2]**)

Lemma 2.5.

$$\sum \frac{\partial^2}{\partial z_j \partial \bar z_k}(\alpha_j \bar\alpha_k e^{-\phi}) = -2\,\mathrm{Re}\,\bar\partial\bar\partial^*_\phi\alpha \cdot \bar\alpha e^{-\phi} + \sum |\frac{\partial \alpha_j}{\partial \bar z_k}|^2 e^{-\phi} \qquad (2.7)$$
$$+ |\bar\partial^*_\phi\alpha|^2 e^{-\phi} - |\bar\partial\alpha|^2 e^{-\phi} + \sum \phi_{j\bar k}\alpha_j\bar\alpha_k e^{-\phi}.$$

If we define a differential form of bidegree $(n-1, n-1)$

$$T_\alpha = \Sigma \alpha_j \bar\alpha_k e^{-\phi} \widehat{dz_j \wedge d\bar z_k}$$

the left hand side of (2.7) is precisely $i\partial\bar\partial T_\alpha$ multiplied by the volume form in $\mathbb{C}^n$. A variant of Lemma 2.6 holds on any Kähler manifold. One should then think of α as a $(n, 1)$-form, define $\gamma = *\alpha$ and put $T_\alpha = \gamma \wedge \bar\gamma$. This was the original formulation of Lemma 2.5 in **[S2]**. For $n = 1$, T_α is just $|\alpha|^2 e^{-\phi}$, and Lemma 2.5 amounts to a computation of the Laplacian of this expression. This is precisely the basic idea of Bochner, whence the term "$\partial\bar\partial$-Bochner–Kodaira method".

We next multiply the identity (2.7) by a positive function w and integrate. To take care of the boundary terms that then appear we need to assume that the form α satisfies the $\bar\partial$-Neumann boundary conditions

$$\alpha_n =: \Sigma \alpha_j \frac{\partial \rho}{\partial z_j} = 0$$

on the boundary of D. Here ρ is a defining function of D so that $D = \{\rho < 0\}$. We then have (see **[B2]** for a proof)

Lemma 2.6. *Assume α is a $(0, 1)$-form with coefficients in $C^2(\bar D)$, satisfying $\alpha_n = 0$ on ∂D. Then*

$$\int \sum \phi_{j\bar k}\alpha_j\bar\alpha_k e^{-\phi} w + \int |\bar\partial^*_\phi \alpha|^2 e^{-\phi} w - \int \sum w_{j\bar k}\alpha_j\bar\alpha_k e^{-\phi}$$
$$+ \int \sum |\bar\partial_k \alpha_j|^2 e^{-\phi} w + \int_\partial w \sum \rho_{j\bar k}\alpha_j\bar\alpha_k e^{-\phi} dS/|\partial\rho| \qquad (2.8)$$
$$= 2\,\mathrm{Re} \int \bar\partial\bar\partial^*_\phi\alpha \cdot \bar\alpha e^{-\phi} w + \int |\bar\partial\alpha|^2 e^{-\phi} w.$$

To obtain Theorem 2.4 from the lemmas we argue as follows. Let u_0 be the minimal solution in $L^2(e^{-\psi})$. Assume f is smooth up to the boundary. From the solution of the $\bar\partial$-Neumann problem it follows that u_0 can be written

$$u_0 = \bar\partial^*_\psi \alpha$$

where α is smooth up to the boundary $\partial\alpha = 0$ and satisfies the $\bar\partial$-Neumann boundary conditions. Lemma 2.6 then gives, if w is plurisuperharmonic and D is pseudoconvex, that

$$\int w\Sigma\psi_{j,k}\alpha_j\bar{\alpha}_k e^{-\psi} + \int w|u_0|^2 e^{-\psi}$$
$$\leq 2\,\mathrm{Re}\int f\cdot\bar{\alpha}e^{-\psi}w$$
$$\leq \int \Sigma\psi^{j,k} f_j \bar{f}_k e^{-\psi} w + \int \Sigma\psi_{j,k}\alpha_j\bar{\alpha}_k e^{-\psi} w.$$

Hence

$$\int w|u_0|^2 e^{-\psi} \leq \int |f|^2_{\partial\bar{\partial}\psi} e^{-\psi} w,$$

and we are done.

In Theorem 2.3 we have made the weight function *less* plurisubharmonic by adding a psh function in the exponent, whereas in Theorem 2.4 we subtract a psh function in the exponent. We next formulate a result from **[Del]** where the function in the exponent only satisfies a first order condition. We use the same notation as in the previous theorems.

Theorem 2.7. *Let ξ be a realvalued function of class C^1 in D which satisfies*

$$i\partial\xi\wedge\bar{\partial}\xi \leq i\partial\bar{\partial}\psi$$

(i.e., the gradient of ξ is uniformly bounded by 1 in the $\partial\bar{\partial}\psi$-metric). Then

$$\int |u_0|^2 e^{\xi-\psi} \leq C\int |f|^2_{\partial\bar{\partial}\psi} e^{\xi-\psi}, \tag{2.9}$$

where C is an absolute constant.

For $n = 1$ an estimate of this type was given, under considerably more general assumptions, in **[Chr]**. That there is for each ξ satisfying the assumption, *some* solution to (2.1) satisfying (2.9), was proved earlier in **[DOh]**. We shall now finally see how one can combine the ideas in Theorems 2.3 and 2.4, to obtain a general boundary value estimate for the canonical solution.

Theorem 2.8. *Let ψ be plurisubharmonic in D and let $w \geq 0$ be either* (i) *plurisuperharmonic in D, or* (ii) *satisfy*

$$i\partial\bar{\partial}w \leq wi\partial\bar{\partial}\psi/2.$$

Let D be given as $D = \{\rho < 0\}$ where ρ is a plurisubharmonic defining function with nonvanishing gradient on the boundary of D. Let u_0 be the minimal solution to (1) *in $L^2(e^{-\psi})$. Then we have in case* (i) *that*

$$\int_{\partial D} |u_0|^2 e^{-\psi} w\,dS/|\partial\rho| \leq \int_D |f|^2_\Theta e^{-\psi} w,$$

with

$$\Theta = i(-\rho\partial\bar\partial\psi + \partial\bar\partial\rho).$$

In case (ii) *we have*

$$\int_{\partial D} |u_0|^2 e^{-\psi} w dS/|\partial\rho| \leq 2\int_D |f|^2_\Theta e^{-\psi} w.$$

If D is e.g. strictly pseudoconvex, it is well known that any continuous and non-negative function w on the boundary of D can be extended to a non-negative function satisfying (i) or (ii) in D. Clearly the optimal extension for our estimate will then be the infimum over all such extensions, i.e., the functions satisfying the Monge–Ampere equations

$$(\partial\bar\partial w)^n = 0 \qquad \text{or} \qquad (\partial\bar\partial w - w\partial\bar\partial\psi/2)^n = 0$$

respectively. In one variable this means that we choose our extension to be either harmonic or satisfying a Schrödinger equation $\Delta w = w\Delta\psi/2$. Still in one variable we can even take limits of continuous functions , and allow our boundary values to be an arbitrary positive measure, e.g. a Dirac measure at a point in the boundary. We then get pointwise control of our solution, i.e., a sup-norm estimate, and we will discuss this more closely in the next section. In higher dimensions however the Monge–Ampere extensions will tend to infinity if we let a sequence of continuous functions tend to a Dirac measure at the boundary. It seems it has not been studied for which measures on the boundary one can find a (suitably defined) extension satisfying a Monge–Ampere equation.

Proof of Theorem 2.8. We prove case (i), the other case being a simple variation on the same theme. Take $\epsilon > 0, r < 1$ and put

$$\phi = r \log 1/(-\rho + \epsilon).$$

A direct computation then shows that

$$i\partial\phi \wedge \bar\partial\phi \leq i(r\partial\bar\partial\phi - r^2\partial\bar\partial\rho/(-\rho + \epsilon)) \tag{2.9}$$

Now, $u_0 e^\phi$ is orthogonal to the space of holomorphic functions in $L^2(e^{-\phi-\psi})$, so we can write

$$u_0 e^\phi = \bar\partial^*_{\phi+\psi}\alpha$$

where α is a $\bar\partial$-closed (0, 1)-form satisfying the $\bar\partial$-Neumann boundary conditions. By the fundamental inequality

$$\int \Sigma(\phi_{j,k} + \psi_{j,k})\alpha_j\bar\alpha_k e^{-\phi-\psi} w + \int |u_0|^2 e^{\phi-\psi} w \leq 2\,\mathrm{Re}\int \bar\partial(u_0 e^\phi)\cdot\bar\alpha e^{\phi-\psi} w$$
$$= 2\,\mathrm{Re}\int f\cdot\bar\alpha e^{-\psi} w + 2\,\mathrm{Re}\int u_0\bar\partial\phi\cdot\bar\alpha e^{-\psi} w,$$

and the last integral is dominated by

$$r \int |u_o|^2 e^{\phi-\psi} w + 1/r \int |\alpha \cdot \bar{\partial}\phi|^2 e^{-\phi-\psi} w.$$

By (2.9)

$$\Sigma \phi_{j,k} \alpha_j \bar{\alpha_k} \geq 1/r |\alpha \cdot \partial \phi|^2 + r \Sigma \rho_{j,k} \alpha_j \bar{\alpha_k}/(-\rho + \epsilon),$$

so altogether we get

$$\int \left(r \Sigma \rho_{j,k} \alpha_j \bar{\alpha_k}/(-\rho+\epsilon) + \Sigma \psi_{j,k} \alpha_j \bar{\alpha_k}\right)(-\rho+\epsilon)^r e^{-\psi} + (1-r) \int |u_0|^2 (-\rho+\epsilon)^{-r} e^{-\psi}$$
$$\leq 2 \operatorname{Re} \int f \cdot \bar{\alpha} e^{-\psi} w.$$

Put

$$\Theta^\epsilon = i((-\rho + \epsilon)\partial\bar{\partial}\psi + \partial\bar{\partial}\rho).$$

We then have

$$r \int \Sigma \Theta^\epsilon_{j,k} \alpha_j \bar{\alpha_k} (-\rho + \epsilon)^{r-1} e^{-\psi} + (1 - r) \int |u_0|^2 (-\rho + \epsilon)^{-r} e^{-\psi}$$
$$\leq 2 \operatorname{Re} \int f \cdot \alpha e^{-\psi} w,$$

from where it follows, first, that

$$(1 - r) \int |u_0|^2 (-\rho + \epsilon)^{-r} e^{-\psi} w \leq 1/r \int |f|^2_{\Theta^\epsilon} (-\rho + \epsilon)^{1-r} e^{-\psi} w,$$

and then, letting ϵ tend to zero,

$$(1 - r) \int |u_0|^2 (-\rho)^{-r} e^{-\psi} w \leq 1/r \int |f|^2_{\Theta} (-\rho)^{1-r} e^{-\psi} w.$$

Finally we can let r tend to 1, and use the fact that for any continuous function, v on $\bar{D}$

$$\lim (1 - r) \int_D v(-\rho)^{-r} = \int_{\partial D} v \, dS/|\partial\rho|.$$

This completes the proof. □

3. A Sobolev estimate

One instance of Theorem 2.3 is when ϕ is of the form

$$\phi = \log(-\rho)$$

where $\rho < 0$ is an exhaustion function of our domain D. In case ρ is plurisubharmonic such a ϕ satisfies condition (2.3) with $r = 1$. Let $\delta(z)$ denote the distance from a point

z in D to the boundary of D. It is a theorem of Diederich and Fornaess, **[DFo]**, that for any smoothly bounded pseudoconvex domain there is a number $\eta > 0$ such that there exists such a plurisubharmonic exhaustion function satisfying $c\delta^{\eta} \leq -\rho \leq C\delta^{\eta}$ for some constants c and C. (Note that the number η can not in general be taken equal to 1.) Taking ψ identically equal to 0 it then follows from Theorem 2.3 (see **[BCh]**) that the L^2-minimal solution to (2.1) satisfies the estimate

$$\int |u_0|^2 \delta^{-r\eta} \leq C_r \int |f|^2 \delta^{-r\eta} \tag{3.1}$$

for some constant C_r for any $r < 1$. Since the Bergman projector, $B(v)$, of a function v can be expressed in terms of the L^2 minimal solution, u_0, to the equation $\bar{\partial}u = \bar{\partial}v$ by $B(v) = v - u_0$ we can obtain from this that the Bergman projector satisfies a similar estimate. Finally one can use the fact that for harmonic functions, the weighted norm in (3.1) is equivalent to the Sobolev norm in $W^{r\eta/2}$ to deduce the following result from **[BCh]** (the same result was obtained by McNeal **[McN2]**).

Theorem 3.1. *Let D be a bounded pseudoconvex domain in $\mathbb{C}^n$ with Lipschitz boundary, such that there exists a plurisubharmonic exhaustion function ρ satisfying $c\delta^{\eta} \leq -\rho \leq C\delta^{\eta}$. Then the Bergman projector is a bounded operator from the Sobolev space W^s to itself for any $s < \eta/2$*

We refer to **[BCh]** for the proof and further results along these lines.

4. Uniform estimates

In this section we will discuss generalizations of the weighted L^2-estimates to uniform norms. We first note that there can be no estimate of the type

$$\sup_D |u|^2 e^{-\phi} \leq C \sup_D |f|^2 e^{-\phi} \tag{4.1}$$

for general plurisubharmonic functions. To see that this is impossible even in one variable, it suffices to take a subharmonic function, ϕ, in D which equals $-\infty$ on a set with a point of accumulation, p, in D. If we choose f to be smooth with support which avoids the singular set of ϕ we can arrange things so that the right hand side of (4.1) is finite. If however the left side is also finite we necessary have that u vanishes on the singular set. Since u would be holomorphic in a neighbourhood of this set it would follow that u vanishes identically on that component of the complement of the support of f that contains p, something which is general impossible to achieve.

Nevertheless we can, in the one-dimensional case, obtain a satisfactory estimate for the boundary values of u (see **[B4]**).

Theorem 4.1. *Let D be a domain in* $\mathbb{C}$ *given as*

$$D = \{\rho < 0\},$$

where ρ *is a subharmonic function in D whose gradient does not vanish on the boundary of D. Let* u_0 *be the solution to* (2.1) *of minimal norm in* $L^2(e^{-\psi})$. *Then*

$$\sup_D |u_0| e^{-\psi/2}/|\partial\rho| \leq \sup_D |f| e^{-\psi/2}/(-\rho\Delta\psi + \Delta\rho). \tag{4.2}$$

An example of this theorem is when D is the disk and $\rho = |z|^2 - 1$. Then the condition that we can find some *bounded* subharmonic ψ such that the right hand side of (4.2) is finite is essentially a Carleson measure condition on $|f|$ (see **[B3]**). Thus we recover the familiar result that we can solve with boundary values in L^∞ for Carleson measure data. The proof of Theorem 4.1 from **[B4]** is based on Kato's inequality and the maximum principle for the Schrödinger operator

$$\Delta u - u\Delta\psi.$$

Another proof (with a slightly worse constant) can be obtained from Theorem 2.8. For this, suppose first the right hand side of (4.2) is bounded by 1. Then Theorem 2.8 implies that

$$\int_{\partial D} |u_0|^2 e^{-\psi} w dS/|\partial\rho| \leq 2\int ((-\rho)\Delta\psi + \Delta\rho)w,$$

if we choose w to be a solution of the equation

$$\Delta w = w\Delta\psi/2.$$

Notice that this equation has a unique solution for any choice of continuous boundary values (**[GT**, Thm. 6.13]), and that the solution will be nonnegative if the boundary values are. Then the right hand side equals

$$2\int 2(-\rho)\Delta w + w\Delta\rho \leq 4\int (-\rho)\Delta w + w\Delta\rho = 4\int w|\partial\rho| dS,$$

where the last equality follows from Green's theorem. Hence we get

$$\int_{\partial D} |u_0|^2 e^{-\psi}/|\partial\rho|^2 w|\partial\rho| dS \leq 4\int w|\partial\rho| dS.$$

Since this holds for any continuous function,w, on the boundary we conclude that

$$\sup_D |u_0|^2 e^{-\psi}/|\partial\rho|^2 \leq 4$$

The fact that interior sup-norm estimates do not hold in one variable implies that such estimates do not even hold on the boundary in higher dimensions for general weights. (Choose ψ and f to be independent of the second variable in, say, the ball in $\mathbb{C}^2$. Then an optimal solution can also be taken independent of the second variable, and a boundary estimate would imply an interior estimate in one variable.) In higher dimensions it therefore becomes inevitable to impose additional conditions on the weight.

Let us first stick to the case of one dimension and ask for which weight functions we have an estimate of the type

$$\sup_D |u|^2 e^{-\psi} \leq C \sup_D |f|^2 e^{-\psi}/\Delta\psi. \tag{4.3}$$

That such an estimate should hold in some cases is not unreasonable if we compare with the L^2-case.

One advantage with this stronger formulation of the problem is that it now becomes invariant under biholomorphic transformations. Moreover, if (4.3) holds for some function ψ, it also holds for $\psi + h$ with h harmonic (at least if D is simply connected, write $h = \operatorname{Re} H$ and substitute $fe^{-H/2}$ for f). The property that (4.3) holds is therefore really a question about $\Delta\psi$. If we associate to ψ the Riemannian metric $ds^2 = \Delta\psi|dz|^2$, and remember that the problem is invariant under conformal transformations, we see that the answer should depend on the curvature properties of ds^2. Here is one simple result in this direction.

Theorem 4.2. *Suppose D is a simply connected domain in $\mathbb{C}$, that ψ is subharmonic in D and that the metric $ds^2 =: \Delta\psi|dz|^2$ has constant Gaussian curvature equal to $-k < 0$. Then there is for any f a solution to* (2.1) *such that* (4.3) *holds with $C = C_k$.*

Proof. Let us first consider the model case when D is the disk and

$$\psi = 1/k \log 1/(1 - |z|^2).$$

The Gaussian curvature then equals

$$-\frac{\Delta \log \Delta\psi}{\Delta\psi} = -k.$$

We want to solve the $\bar\partial$-equation with an estimate

$$\sup |u|^2(1 - |z|^2)^{1/k} \leq C \sup |f|^2(1 - |z|^2)^{1/k+2}$$

in this case. This is easily done using the integral formula

$$u(z) = 1/\pi \int \left(\frac{1 - |\zeta|^2}{1 - \bar\zeta z}\right)^\tau \frac{f(\zeta)}{\zeta - z},$$

with $\tau > 1/2k$. This proves the theorem in the model case. In general we can find an isometry between D with the metric ds^2 and a subdomain of the disk with the metric

$$\Delta 1/k \log 1/(1 - |z|^2)|dz|^2.$$

Since we may choose the isometry orientation preserving and thus conformal, and since the estimate is conformally invariant, we may assume that D is that subdomain.

Extending f trivially by putting it equal to zero in the rest of the disk, we are back in the model case and the theorem follows. □

We finally turn to the case of higher dimensions. First we state a result from (**[B2]**).

Theorem 4.3. *Let $D = B$ be the unit ball in $\mathbb{C}^n$ and let ψ be plurisubharmonic in B. Let*

$$\omega = i\partial\bar{\partial}\log 1/(1-|z|^2)$$

be the Kähler form of the Bergman metric and assume that

$$N\omega \leq i\partial\bar{\partial}\psi \leq M\omega$$

where $N > (n+1)^2$. Let finally u_0 be the $L^2(e^{-\psi})$-minimal solution to (2.1). *Then*

$$\sup|u_0|^2 e^{-\psi} \leq C_{N,M}\sup|f|^2_{\partial\bar{\partial}\psi}e^{-\psi}.$$

Of course we could as well have written $|f|^2_\omega$ in the conclusion of the theorem, since it is a part of the assumption that the two norms are comparable, but we have written $\partial\bar{\partial}\psi$ to emphasize the relation to Theorem 4.2. A somewhat weaker estimate is also valid without the assumption $i\partial\bar{\partial}\psi \leq M\omega$; see **[B2]** for this, but at present the lower bound in the hypothesis on $i\partial\bar{\partial}\psi$ is necessary for the proof to work.

Of special interest is the case of bounded weight functions. Then the weighted estimates are of course equivalent to unweighted estimates, but we may still gain something by measuring the norms with respect to the $\partial\bar{\partial}\psi$-metric (for a variant of this compare the discussion after Theorem 4.1.

Definition 4.4. Let f be a form in a domain D. We say that f is quasibounded if there exists a bounded plurisubharmonic function, ϕ , in D such that $|f|_{\partial\bar{\partial}\phi}$ is uniformly bounded in D.

Notice that if ϕ is uniformly bounded, then $\exp(\phi)$ will satisfy (2.3). Hence, if f is quasibounded there exists in particular a function $\tilde{\phi}$ satisfying (2.3), such that $|f|_{\partial\bar{\partial}\tilde{\phi}}$ is uniformly bounded. Being quasibounded is however a stronger property (e.g. a form that is bounded with respect to the Bergman metric in the ball is not necessarily quasibounded).

It is not true that we can solve $\bar{\partial}u = f$ with u uniformly bounded if f is quasibounded, but the following conjecture seems plausible.

Conjecture 1. Let $D = B$ be the ball in $\mathbb{C}^n$ and let f be a $\bar{\partial}$-closed (0, 1)-form in B, such that f and $\bar{\partial}f$ are both quasibounded. Then there is a bounded function u such that $\bar{\partial}u = f$.

The main interest of the conjecture is that it would imply the corona theorem in the ball, and the conjecture does hold for $n = 1$. It is then essentially a reformulation of a result of T. Wolff, see **[B2]** for a discussion of this.

We end with a result showing that it is quite easy to prove the conjecture "within ϵ".

Theorem 4.5. *Let $D = B$ be the unit ball in $\mathbb{C}^n$, and let u_0 be the L^2-minimal solution to* (2.1). *Assume there are constants C and $\eta > 0$ such that*

$$|f|^2_{\partial\bar\partial\phi} \leq C(1-|z|^2)^\eta,$$

and

$$|f|^2_\omega \leq C,$$

where ϕ satisfies (2.3) *and ω is the Kähler form of the Bergman metric. Then u_0 is bounded in B.*

Notice that the second assumption on f, that f be bounded with respect to the Bergman metric in itself is not sufficient to get a bounded solution (take $f = 1/(1-|z|)$ in the disk).

Proof. For a function v in L^2 we let $\pi(v)$ denote the orthogonal projection of v to the space of functions that are orthogonal to the Bergman space of holomorphic functions in L^2. Explicitly

$$\pi(v)(z) = v(z) - c_n \int \frac{v(\zeta)}{(1-\bar\zeta\cdot z)^{n+1}}.$$

The crucial observation is now that since u_0 is orthogonal to the Bergman space we have

$$\int u_0\bar v = \int u_0\overline{\pi(v)} = \int u\overline{\pi(v)}$$

if u is *any* solution to (2.1) in L^2. Applying Theorem 2.1 we find that

$$\left|\int u_0\bar v\right|^2 \leq \int |u|^2 e^{-\psi}\int |\pi(v)|^2 e^{\psi} \leq \int |f|^2_{\partial\bar\partial\phi} e^{-\psi}\int |\pi(v)|^2 e^{\psi},$$

if ψ is plurisubharmonic. Let a be an arbitrary point in B and let

$$K_a = \left\{z;\ \frac{(1-|z|^2)(1-|a|^2)}{|1-z\cdot\bar a|^2} > 1/2\right\},$$

i.e., the image of the ball with radius $1/\sqrt{2}$ under a biholomorphism that sends 0 to a. Put $v = u_0\chi_{K_a}$. A simple computation, which we spare the reader, shows that

$$|\pi(v)|^2(\zeta) \leq C\left(|u_o|^2\chi_{K_a} + \frac{(1-|a|^2)^{n+1}}{|1-\bar a\cdot\zeta|^{2n+2}}\int_{K_a}|u_0|^2\right).$$

Now take

$$\psi = \tau\log|1-\bar\zeta\cdot a|,$$

with $n+1<\tau<n+1+\eta$. It is then easily verified that the hypothesis $|f|^2_{\partial\bar\partial\phi} \leq C(1-|z|^2)^\eta$ implies that

$$\int |f|^2_{\partial\bar\partial\phi} e^{-\psi} \leq C,$$

and moreover

$$\int |\pi(v)|^2 e^{\psi} \leq (1-|a|^2)^{n+1} \int_{K_a} |u_0|^2,$$

so all in all we obtain

$$(1-|a|^2)^{-n-1} \int_{K_a} |u_0|^2 \leq C.$$

This means that the average of $|u_0|^2$ over K_a is bounded independently of a. Invoking the hypothesis $|f|^2_\omega \leq C$ we find that u_0 is itself uniformly bounded, and the proof is complete. □

The fact that makes the proof work is that we have explicit knowledge of the Bergman kernel in the ball. For this reason it is not clear how to generalize the estimate to e.g. the minimal solution in weighted L^2-space. On the other hand, the proof should generalize to other situations where we have size estimates of the Bergman kernel, e.g. 2-dimensional or convex domains of finite type.

References

[B1] Berndtsson, B., Some recent results on estimates for the $\bar\partial$-equation, in: Contributions to complex analysis and analytic geometry, Aspects Math. E 26, Vieweg, Braunschweig 1994, 27–42.

[B2] —, Uniform estimates with weights for the $\bar\partial$-equation, J. Geom. Anal. 7 (1997), 195–215.

[B3] —, $\bar\partial_b$ and Carleson type inequalities, in: Complex Analysis II, College Park, MD, 1985–86, (Berenstein, C. A., ed.), Lecture Notes in Math. 1276, Springer-Verlag, Berlin–New York, 1987, 42–54.

[B4] —, $\bar\partial$ and Schrödinger operators, Math. Z. 221 (1996), 401–413.

[BCh] Berndtsson, B., and Charpentier, Ph., A Sobolev mapping property of the Bergman kernel, Math. Z. 235 (2000), 1–10.

[Chr] Christ, M., On the $\bar\partial$ equation in weighted L^2 norms in C^1, J. Geom. Anal. 1 (1991), 193–230.

[Del] Delin, H., Pointwise estimates for the weighted Bergman projection kernel in $\mathbf{C}^n$, using a weighted L^2 estimate for the $\bar\partial$ equation, Ann. Inst. Fourier (Grenoble) 48 (1998), 967–997.

[Dem] Demailly, J.-P., Estimations L^2 pour l'opérateur $\bar{\partial}$ d'un fibre vectoriel holomorphe semi-positif au-dessus d'une varieté Kählerienne complète, Ann. Sci. École Norm. Sup (4) 15 (1982), 457–511.

[DF] Donnelly, H., and Fefferman, C., L^2-cohomology and index theorem for the Bergman metric, Ann. of Math. (2) 118 (1983), 593–618.

[DFo] Diederich, K., and Fornaess, J., Pseudoconvex domains: bounded strictly plurisubharmonic exhaustion functions, Invent. Math. 39 (1977), 129–141.

[DH] Diederich, K., and Herbort, G., Extension of holomorphic L^2-functions with weighted growth conditions, Nagoya Math. J. 126 (1992), 141–157.

[DOh] Diederich, K., and Ohsawa T., An estimate for the Bergman distance on pseudoconvex domains, Ann. of Math. (2) 141 (1995), 181–190.

[GT] Gilbarg, D., and Trudinger, N., Elliptic partial differential equations of second order (second edition), Grundlehren Math. Wiss. 224, Springer-Verlag, Berlin–New York 1983.

[H] Hörmander , L., An introduction to complex analysis in several variables, D. Van Nostrand Co., Inc., Princeton (NJ)–Toronto (Ont.)–London 1966.

[McN1] McNeal, J., On large values of L^2 holomorphic functions, Math. Res. Lett. 3 (1996), 247–259.

[McN2] —, Private communication.

[McN3] Invariant metric estimates for $\bar{\partial}$ on some pseudoconvex domains, Ark. Mat., to appear.

[OhT] Ohsawa, T., On the extension of L^2 holomorphic functions, Math. Z. 195 (1987), 197–204.

[Oh] Ohsawa, T., Recent applications of L^2 estimates for the operator $\bar{\partial}$, in: Proceedings of the International Congress of Mathematicians, Vol. I, II (Kyoto, 1990), Math. Soc. Japan, Tokyo 1991, 913–921.

[S1] Siu, Y-T., The Fujita conjecture and the extension theorem of Ohsawa–Takegoshi, in: Geometric complex analysis (Hayama, 1995), World Sci. Publishing, River Edge, NJ, 1996, 577–592.

[S2] —, Complex-analyticity of harmonic maps, vanishing and Lefschetz theorems, J. Differential Geom. 17 (1982), 55–138.

Department of Mathematics
CTH S-412 96
Göteborg, Sweden
bobmath.chalmers.se

Hypoellipticity in the infinitely degenerate regime

*Michael Christ**

Contents

Let $\{X_j\}$ be a collection of real vector fields with C^∞ coefficients, defined in a neighborhood of a point $x_0 \in \mathbb{R}^d$. Consider a second order differential operator $L = -\sum_j X_j^2 + \sum_j \alpha_j X_j + \beta$ where α_j, β are C^∞ real coefficients. A well known sufficient condition **[17]** for L to be C^∞ hypoelliptic is that the Lie algebra generated by $\{X_j\}$ should span the tangent space to $\mathbb{R}^d$ at x_0. This bracket condition is by no means necessary; no satisfactory characterization of hypoellipticity exists, and it appears unlikely that one could be found. The purposes of this note are:

1. To establish sufficient conditions for hypoellipticity for operators such as $-\sum_j X_j^2$, for the Kohn Laplacian on pseudoconvex three-dimensional CR manifolds, and for the $\bar\partial$-Neumann problem in $\mathbb{C}^2$, in the case of infinite type.

2. To point out an inequality weaker than the subelliptic estimates which implies hypoellipticity, and which is the weakest possible such inequality.

*Research supported by NSF grant DMS96-23007. This paper was completed while the author was on appointment as a Miller Research Professor in the Miller Institute for Basic Research in Science.

3. To popularize within the $\bar{\partial}$-Neumann community certain ideas developed in another context.

4. To emphasize the parallel between the theories of C^∞ and analytic/Gevrey class hypoellipticity.

This paper has undergone several revisions since its first version, written in the Spring of 1996. Although the results and methods employed here are new within the context of the $\bar{\partial}$-Neumann problem, I have subsequently learned that when viewed in the wider context of sums of squares of vector fields and related operators, they substantially overlap works of other authors, some earlier and some contemporaneous, including but not limited to **[19]**, **[26]**, **[27]**; other selected relevant works are listed in the bibliography. Thus this paper is in part expository.

It has subsequently been shown by Kohn **[20]** how some of these results may be extended to $\Box_b$ and the $\bar{\partial}$-Neumann problem in higher dimensions.

1. Examples

To orient the reader we review a few concrete examples of operators which need not satisfy the bracket hypothesis. Consider three classes of operators:

$$\begin{aligned} L_1 &= -\partial_x^2 - a^2(x)\partial_t^2 && \text{in } \mathbb{R}^2 \\ L_2 &= -\partial_x^2 - a^2(x)\partial_t^2 - \partial_y^2 && \text{in } \mathbb{R}^3 \\ L_3 &= -\partial_x^2 - a^2(x)\partial_t^2 - b^2(x)\partial_y^2 && \text{in } \mathbb{R}^3 \end{aligned}$$

Assume that $a, b \in C^\infty$, that a, b are even and nonnegative, and are nondecreasing on $[0, \infty)$, and that $a(x) = 0 \Leftrightarrow x = 0$. Concerning operators L_3, assume that $b(x) \geq a(x)$ for all x, and likewise that $b(x) = 0 \Leftrightarrow x = 0$.

Proposition 1.1 (Fediĭ **[13]**). *Under the above hypotheses, all operators L_1 are hypoelliptic.*

Proposition 1.2 (Kusuoka and Stroock **[21]**). *L_2 is hypoelliptic if and only if*

$$\lim_{x\to 0} x \log a(x) = 0. \tag{1.1}$$

This was originally proved by stochastic techniques; see Theorem 8.41 of **[21]**. To indicate how these examples fit within the general context discussed in this paper, we will indicate alternative proofs of Propositions 1.1 and 1.2.

Proposition 1.3. *Suppose that the coefficients $a, b \in C^\infty$ vanish where $x = 0$ but nowhere else, and that $b(x) \geq a(x) \geq 0$ for all x. If* $\limsup_{x\to 0} |x \log a(x)| \neq 0$, *and if the coefficient b satisfies*

$$\lim_{x\to 0} b(x)x|\log a(x)| = 0, \tag{1.2}$$

then L_3 is hypoelliptic. Moreover, if some partial derivative of b is nonzero at $x = 0$, then L_3 is hypoelliptic if and only if (1.2) *is satisfied.*

This result was discovered independently by other authors. What is interesting about this example is that increasing the size of b makes the operator L_3 stronger in the sense that the associated quadratic form $\langle L_3 f, f\rangle$ becomes larger for every f, but makes L_3 less likely to be hypoelliptic. In the context of analytic hypoellipticity such a phenomenon has already been observed: Métivier **[23]** has shown that $\partial_x^2 + (x\partial_t)^2 + (t\partial_t)^2$ is not analytic hypoelliptic, while the weaker operator $\partial_x^2 + (x\partial_t)^2$ is analytic hypoelliptic **[15]**. A parallel result for Gevrey hypoellipticity is that when $a(x) = x^q$ and $b(x) = x^p$ with $q \geq p \geq 1$, L_3 is hypoelliptic in the Gevrey class of order s if and only if $b \cdot a^{-1/s}$ is bounded as $x \to 0$ **[7]**; see also **[5]** and **[24]**.

The next proposition is a special case of Theorem 2.5.3 of Oleĭnik and Radkevič[1] **[28]**.

Proposition 1.4. *Suppose that $L = -\sum_j X_j^2$ where the X_j are smooth real vector fields in some open set V. Suppose that at each point of V, at least one X_j is nonzero. Suppose that $\{X_j\}$ satisfies the bracket hypothesis at all but finitely many points of V. Then L is hypoelliptic in V.*

The analogue for C^ω hypoellipticity is false; $\partial_x^2 + (x\partial_t)^2 + (t\partial_t)^2$ is elliptic except at a single point, yet is not analytic hypoelliptic **[23]**. For other such examples in the analytic and Gevrey class contexts see **[8]**, **[9]**.

2. Main results

Our results are most naturally formulated in a somewhat more general framework. Denote by $S^m_{\rho,\delta}$ the usual classes of symbols for pseudodifferential operators **[31]**, **[32]**. Denote by $S^{m+}_{1,0}$ the intersection, over all $\varepsilon > 0$, of all classes $S^{m+\varepsilon}_{1-\varepsilon,\varepsilon}$. Given any class S of symbols, we denote by $\mathrm{Op}(S)$ the associated class of pseudodifferential operators, with respect to the quantization of the Kohn–Nirenberg calculus. In particular we consider classes $\mathrm{Op}(S^m_{\rho,\delta})$, $\mathrm{Op}(S^{m+}_{1,0})$ of operators.

All functions, distributions and symbols are permitted to be complex vector valued except where otherwise noted; $\langle u, v\rangle$ denotes either a Hermitian pairing of square integrable vector valued functions, or of test functions with distributions. The symbol $\|u\|$, with no subscript, denotes the L^2 norm. For any open subset $V \subset \mathbb{R}^d$, we

[1]The theorem cited concerns *global* hypoellipticity, but for operators hypoelliptic in $U\backslash\{x_0\}$, hypoellipticity and global hypoellipticity in U are equivalent.

denote by T^*V the complement of the zero section in the cotangent bundle of V. X^* denotes the formal adjoint of any operator X. $C_0^k(V)$ and $C_0^\infty(V)$ denote the classes of functions compactly supported in V and belonging to C^k, C^∞ respectively.

Fix an open neighborhood V of x_0, and suppose L to be a linear operator mapping $\mathcal{D}(V)$ to $\mathcal{D}'(V)$ taking the form

$$L = \sum_j X_j^* X_j + \sum_j A_j X_j + \sum_j X_j^* \tilde{A}_j + A_0 \tag{2.1}$$

where $X_j \in \mathrm{Op}(S^1_{1,0})$ and $A_j, \tilde{A}_j, A_0 \in \mathrm{Op}(S^0_{1,0})$. This sum and similar ones will always be taken over all $1 \le j \le J$ for some unspecified integer J. More generally, we consider any $L \in S^2_{1,0}$ whose full symbol equals the full symbol of such an operator, in some conic subset $\Gamma \subset T^*V$.

A fundamental example is the Kohn Laplacian $\Box_b = \bar\partial_b^* \bar\partial_b + \bar\partial_b \bar\partial_b^*$ defined on (p, q) forms on any CR manifold; this is a second order system of partial differential operators with complex coefficients.[2] A closely related second example (up to composition with a harmless elliptic factor) is the pseudodifferential Calderón operator for any smoothly bounded domain in $\mathbb{C}^2$, arising from application of the boundary reduction method to the $\bar\partial$-Neumann problem.

Write $\langle\xi\rangle = (e^2 + |\xi|^2)^{1/2}$ for any $\xi \in \mathbb{R}^d$. By $\log^2(y)$ we mean $(\log y)^2$. By a conic open subset of $T^*\mathbb{R}^d$ we will always mean such a subset disjoint from the zero section. Denote by H^s the usual Sobolev space of order s, and by $WF_{H^s}(u)$ the H^s wave front set of a distribution u. Denote by $\hat{u}$ the Fourier transform of u.

All operators studied in this paper will satisfy global inequalities of the form

$$\int_{\mathbb{R}^d} w^2(\xi)|\hat{u}(\xi)|^2\, d\xi \le C \sum_j \|X_j u\|^2 + C\|u\|^2 \ \text{ for all } u \in C_0^1(V), \tag{2.2}$$

where V is some fixed open set in which hypoellipticity is to be studied, $C < \infty$ is a fixed constant, and w is a strictly positive, continuous function satisfying

$$w(\xi) \to \infty \ \text{ as } |\xi| \to \infty. \tag{2.3}$$

This will either be assumed explicitly, or will be a consequence of other hypotheses.

Theorem 2.1. *Let L take the form* (2.1). *Suppose that there exists a function w satisfying*

$$\frac{w(\xi)}{\log\langle\xi\rangle} \to \infty \ \text{ as } |\xi| \to \infty \tag{2.4}$$

for which (2.2) *holds. Then L is hypoelliptic in V. More precisely, for any $s \in \mathbb{R}$ and $u \in \mathcal{D}'(V)$,*

$$WF_{H^s}(u) \subset WF_{H^s}(Lu). \tag{2.5}$$

[2] It is not identical to the operator referred to by the same name in **[22]**.

The hypothesis (2.4) is the optimal condition of its type. For instance:

Proposition 2.2. *Operators of the type L_2 discussed in Proposition* 1.2 *are hypoelliptic if and only if they satisfy* (2.4).

Moreover, the operator $-\partial_{x_1}^2 - \partial_{x_2}^2 - e^{-2/|x_1|}\partial_{x_3}^2$ in $\mathbb{R}^3$ satisfies the inequality (2.2) with $w(\xi) = \log\langle\xi\rangle$, and fails to be hypoelliptic. See §5.

Theorem 2.1, and the observation that (2.4) is necessary for certain operators, are due originally to Y. Morimoto. See for instance **[25]**, **[37]**, **[26]**.

Nonetheless, it is important to understand that (2.4) is very far from being a necessary condition for hypoellipticity. Consider for instance operators of the type L_1, with $a(x) = 0$ if and only if $x = 0$. These all satisfy the compactness inequality (2.2) for some weight w tending to ∞, but w may tend to infinity arbitrarily slowly; like operators L_2, they satisfy (2.4) if and only if $x \log a(x) \to 0$ as $x \to 0$. Yet they are all hypoelliptic. The same remarks apply to the operators with isolated degeneracies described in Proposition 1.4.

An equivalent formulation of (2.4) is that for each $\delta > 0$ there should exist $C_\delta < \infty$ such that for each real valued function $u \in C_0^2(V)$,

$$\int_{\mathbb{R}^d} \log^2\langle\xi\rangle\, |\hat{u}(\xi)|^2\, d\xi \le \delta \sum_j \|X_j u\|^2 + C_\delta \|u\|^2.$$

Further definitions are required in order to formulate our main result. A point x_0 is said to belong to the complement of the H^s singular support of a distribution u if there exists a distribution $v \in H^s$ such that $v \equiv u$ in some neighborhood of x_0. For any conic subset $R \subset T^*\mathbb{R}^d$, write $u \in H^s(R)$ to mean that $WF_{H^s}(u) \cap R = \emptyset$. Similarly $u \in C^\infty(R)$ means that $WF(u) \cap R = \emptyset$.

A set of vector fields $\{X_j\}$ is said to satisfy the microlocal bracket hypothesis at a point (x_0, ξ_0) of the unit cosphere bundle S^*V if some iterated Poisson bracket

$$\{\sigma_1(X_{j_1}), \{\sigma_1(X_{j_2}), \ldots\}\ldots\}(x_0, \xi_0)$$

of their principal symbols $\sigma_1(X_j)$ is nonzero **[4]**. Write $X_j = \mathrm{Op}(\sigma(X_j))$. Recall that the Poisson bracket of two functions $f, g \in C^1(T^*\mathbb{R}^d)$ is defined to be

$$\{f, g\}(x, \xi) = \sum_{j=1}^d \frac{\partial f}{\partial x_j}\frac{\partial g}{\partial \xi_j} - \frac{\partial g}{\partial x_j}\frac{\partial f}{\partial \xi_j}. \tag{2.6}$$

The principle underlying our analysis (when L is a sum of squares of vector fields X_j) is that hypoellipticity is governed by a semiglobal comparison of (i) the Hamiltonian vector fields associated to the principal symbols of $\{X_j\}$, with (ii) the size of those principal symbols. The larger the latter, the more likely is an operator to be hypoelliptic; the larger the former, the less likely. The following result and its variant Theorem 2.4, formulated below, express this principle. Together they subsume all other sufficient conditions for hypoellipticity formulated in this paper.

Main Theorem 2.3. *Let $R \subset T^*V$ be any ray. Assume that L takes the form* (2.1) *in some conic neighborhood of R, and satisfies* (2.2) *for some $w \in C^\infty$ such that $w(\xi) \to \infty$ as $|\xi| \to \infty$.*

Suppose that for each small conic neighborhood Γ of R there exist scalar valued symbols $\psi, p \in S^0_{1,0}$ such that ψ is everywhere nonnegative, $\psi \equiv 0$ in some smaller conic neighborhood of R, $\psi \geq 1$ on $T^*V \backslash \Gamma$, $p \equiv 0$ in a conic neighborhood of the closure of Γ, and such that for each $\delta > 0$ there exists $C_\delta < \infty$ such that for any relatively compact open subset $U \Subset V$ and for all $u \in C^2_0(U)$ and each index i,

$$\| \operatorname{Op}\big[\log\langle\xi\rangle\{\psi, \sigma(X_i)\}\big]u\|^2 \leq \delta \sum_j \|X_j u\|^2 + C_\delta \|u\|^2 + C_\delta \| \operatorname{Op}(p)u\|^2_{H^1}. \tag{2.7}$$

Then for any $u \in \mathcal{D}'(V)$, $Lu \in C^\infty(R) \Rightarrow u \in C^\infty(R)$. Likewise for each $s \in \mathbb{R}$ and any u, $Lu \in H^s(R) \Rightarrow u \in H^s(R)$.

Because $\{\psi, \sigma(X_i)\} \in S^0_{1,0}$, the symbol of $\operatorname{Op}\big[\log\langle\xi\rangle\{\psi, \sigma(X_i)\}\big]$ is bounded by $\log\langle\xi\rangle$. Hence Theorem 2.3 directly implies Theorem 2.1, by Gårding's inequality and pseudodifferential calculus.

Two points require clarification, however. First, all norms are taken over V, and the constants C_δ are permitted to depend on U. Note that because all pseudodifferential operators occurring here are pseudolocal, an equivalent statement is obtained by taking the L^2 and H^1 norms over an arbitrarily small neighborhood of the closure of U, rather than over V. Second, L is assumed to take the form (2.1) only in a small conic neighborhood of R; the symbols of the operators $X_j \in \operatorname{Op}(S^1_{1,0})$ are extended arbitrarily so that those operators become globally defined. The conclusion (2.7) is independent of the choices of extensions, because of the presence of the term involving $\operatorname{Op}(p)u$ on the right hand side.

The main hypothesis is the existence of ψ having a favorable commutation relation with each X_i. The symbol p plays a subsidiary role; its presence on the right hand side of the inequality means that commutation with ψ need be controlled only near R.

We do not believe that there can be any simple characterization on the level of symbols of those operators satisfying the hypotheses of Theorem 2.3, analogous to the bracket hypothesis for subellipticity; this is a substantial weakness in the theory. It is a more difficult problem to characterize pairs of pseudodifferential operators $A, B \in S^1_{1,0}$ for which B dominates A, in the sense that $\|Au\| \leq C\|Bu\| + C\|u\|$ for all functions u, for general B than for subelliptic B, because the connection between the strength of an operator and the size of its symbol becomes more tenuous when the symbol is permitted to vanish somewhere to infinite order **[14]**.

The next variant has weaker hypotheses and conclusion. An application will be given below in Theorem 3.5.

Theorem 2.4. *Let L take the form* (2.1) *in some open set $V \subset \mathbb{R}^d$, and let $x_0 \in V$. Suppose that for each neighborhood $U \Subset V$ of x_0 there exist scalar valued functions $\Psi \in C^\infty(V)$ and $\eta \in C^\infty(V)$ such that $\Psi \geq 0$, $\Psi \equiv 0$ in a neighborhood of x_0,*

$\Psi \geq c > 0$ on $V \setminus U$, and $\Psi > 0$ on a neighborhood of the support of η. Suppose further that for every $\delta > 0$ there exists $C_\delta < \infty$ such that $\psi(x, \xi) = \Psi(x)$ satisfies

$$\| \operatorname{Op}\big[\log\langle\xi\rangle\{\psi, \sigma(X_i)\}\big]u\|^2 \leq \delta \sum_j \|X_j u\|^2 + C_\delta \|u\|^2 + C_\delta \|\eta u\|^2_{H^1}, \tag{2.8}$$

uniformly for all $u \in C_0^\infty(U)$. Then for any $u \in \mathcal{D}'(V)$,

$$x_0 \notin \text{singular support}\,(Lu) \Rightarrow x_0 \notin \text{singular support}\,(u).$$

Moreover for any $s \in \mathbb{R}$, if Lu belongs to H^s in some neighborhood of x_0, then so does u.

Our analysis is based on conjugation with pseudodifferential operators of variable order. It is related to certain variants[3] of the FBI transform used to study Gevrey class hypoellipticity in **[7]**. The use of operators having favorable commutation relations with a given family of vector fields is a common technique, used for example in the work of Sjöstrand **[29]**, **[30]** and of Tartakoff and other authors on analytic hypoellipticity, and of Boas and Straube **[3]** on global C^∞ regularity in the $\bar\partial$-Neumann problem. The same, or closely related, methods are employed in **[25]**, **[19]**, **[26]**, **[27]**, and other works.

3. Some applications

The motivation for this work was fourfold. First, a theorem of Derridj and Zuily **[11]** (together with its proof) asserts that for vector fields with real analytic coefficients[4], a subelliptic estimate of order ε implies that $\sum X_j^2$ is hypoelliptic in the Gevrey class G^s for all $s \geq \varepsilon^{-1}$; this exponent s is optimal, in general. Since $\varepsilon \to 0$ as $s \to \infty$, and since $G^s \subset C^\infty$ for all s, this suggested that some limiting bound weaker than subellipticity should suffice for C^∞ hypoellipticity.[5] Theorem 2.1 confirms this idea.

A second motivation was work of Bell and Mohammed **[2]**, Kusuoka and Stroock **[21]** and Malliavin **[22]** establishing hypoellipticity in certain nonsubelliptic cases by stochastic methods. Consider any finite collection $\{X_j\}$ of C^∞ vector fields, defined in a neighborhood V of $x_0 \in \mathbb{R}^d$. For $k \geq 1$ denote by $\mathfrak{g}_k$ the $C^\infty(\mathbb{R}^d)$ module spanned by all Lie brackets of the X_j having less than or equal to k factors. Assume:

1. $M \subset V$ is a C^∞ hypersurface.

[3]These variants take the form $\mathcal{F}u(x, \xi) = \int \exp(i(x - y) \cdot \xi - N\psi(x, \xi) \log\langle\xi\rangle)\, u(y)\eta(y)\, dy$, where $\eta \in C_0^\infty$ is a cutoff function, N is a large parameter, and $\psi \in C_0^1$ is an arbitrary nonnegative symbol.

[4]It suffices that the coefficients belong to G^s.

[5]The analogy between the hypotheses of Derridj and Zuily and those of Theorem (2.1) is imperfect; inequality (2.2) with $w(\xi) = \langle\xi\rangle^\varepsilon$ suffices for $G^{1/\varepsilon}$ hypoellipticity, while $w(\xi) = \log\langle\xi\rangle$ is not quite sufficient for C^∞ hypoellipticity. This has to do with the different natures of the topologies for G^s and for C^∞, the former being a countable union of Banach spaces, the latter a countable intersection.

2. There exist k and a collection of vector fields $Z_{i,\alpha} \in \mathfrak{g}_k$, where $1 \le i \le d$ and α ranges over some finite index set, which spans the tangent space to $\mathbb{R}^d$ at each point of $V \backslash M$.

3. At each point of M, at least one vector field X_j is transverse to M.

4. The coefficient

$$\beta(x) = \sum_{\alpha} \left(\text{determinant}(Z_{i,\alpha})_{1 \le i \le d} \right)^2 (x) \tag{3.1}$$

degenerates weakly as $x \to M$ in the sense that

$$\lim_{V \ni x \to M} \text{distance}\,(x, M) \cdot |\log \beta(x)| = 0. \tag{3.2}$$

The theorem of Bell and Mohammed asserts that a slightly stronger version of these hypotheses implies hypoellipticity.[6]

Lemma 3.1. *Let $\{X_j\}$ satisfy the hypotheses enumerated above. Then for every relatively compact open subset $U \Subset V$ and each small $\delta > 0$ there exists $C_\delta < \infty$ such that for all $u \in C_0^\infty(U)$,*

$$\int \log^2 \langle \xi \rangle |\hat{u}(\xi)|^2 \, d\xi \le \delta \sum_j \|X_j u\|^2 + C_\delta \|u\|^2. \tag{3.3}$$

Corollary 3.2. *If a collection $\{X_j\}$ of vector fields satisfies the hypotheses enumerated above, then $L = -\sum_j X_j^2$ is hypoelliptic.*

I emphasize that these first two motivations have retroactively proved spurious, in the sense that Theorem 2.1 was already known **[25]** to other authors.

A third motivation was the search for sufficient conditions for hypoellipticity of the $\bar{\partial}$-Neumann problem for smoothly bounded pseudoconvex domains of infinite type, a topic closely linked to the theory of sums of squares of real vector fields, but calling for generalizations involving complex vector fields, pseudodifferential operators, and systems, and thus perhaps not amenable to stochastic methods. The next result is a direct consequence of Theorem 2.3, Lemma 3.1, and the method of reduction to the boundary as in **[6]**.[7]

Corollary 3.3. *Let Ω be a smoothly bounded pseudoconvex relatively compact open subset of $\mathbb{C}^2$. Suppose that near $x_0 \in \partial\Omega$, the set of weakly pseudoconvex points*

[6]In **[2]** it was assumed that $\beta(x) \ge \exp(-\,\text{distance}\,(x, M)^{-p})$ as $x \to M$, for some $p < 1$. Morimoto **[25]** had earlier established hypoellipticity under the sharp hypothesis distance $(x, M) \cdot |\log \beta(x)| \to 0$, in the special case $k = 1$ where all $Z_{i,\alpha}$ in hypothesis (3.1) belong to the span of the X_j.

[7]Our technique appears applicable to some extent in $\mathbb{C}^d$ for $d > 2$, but we have not carried out the details of the boundary reduction. The formulation of Corollary 3.3 would require certain modifications; for instance, for hypoellipticity on the level of (0, 1) forms, M should be assumed to have real dimension d, and to be totally real.

of $\partial\Omega$ *is contained in a smooth hypersurface*[8] M*, which is everywhere transverse to* $T^{1,0} \oplus T^{0,1}(\partial\Omega)$*. Suppose also that the Levi form satisfies*

$$\lim_{\partial\Omega \ni x \to M} \text{distance}\,(x, M) \cdot \log \lambda(x) = 0.$$

Then the $\bar\partial$*-Neumann problem for* $(0, 1)$ *forms on* Ω *is hypoelliptic in a neighborhood of* x_0.

Here the Levi form is identified with a real-valued function.

This result is derived from the following consequence of Theorem 2.1.

Corollary 3.4. *Let* $\Omega \subset \mathbb{C}^2$ *be a bounded open set with* C^∞ *boundary. Suppose that there exists a symbol* $\mu \in S^1_{1,0}(\partial\Omega)$ *which in any local coordinate system is real and nonnegative modulo addition of a symbol in* $S^0_{1,0}$ *and satisfies*

$$\mu(x, \xi)/\log\langle\xi\rangle \to \infty \quad as\ \langle\xi\rangle \to \infty,$$

such that for every $u \in C^1_0(\partial\Omega)$,

$$\|\operatorname{Op}(\mu)u\| \le C\|\bar\partial_b u\|_{L^2} + C\|\bar\partial_b^* u\| + C\|u\|_{L^2}$$

for some $C < \infty$*. Then the* $\bar\partial$*-Neumann problem for* Ω *is* C^∞ *hypoelliptic.*

J. J. Kohn has asked whether hypoellipticity always holds in this context, provided that the set of weakly pseudoconvex points consists of a smooth real curve that is everywhere transverse to the complex tangent space, but without any restriction on the rate at which the Levi form degenerates. In a forthcoming paper we will show that this is not the case, even though quite weak supplemental hypotheses can suffice to imply hypoellipticity.

A fourth motivation was the hope that a more penetrating study of C^∞ hypoellipticity might shed light on the more intrinsically interesting C^ω case. The present paper achieves for the C^∞ theory a rough parity with what is presently known, in the positive direction, about analytic hypoellipticity.[9]

We now formulate a few variants of the main results. The next theorem and corollary are rudimentary results, in which the symbols ψ required for the application of Theorems 2.3 and 2.4 may be constructed so that their gradients are supported in regions where L is elliptic, or at least is subelliptic. The commutator bounds (2.7) and (2.8) then hold with a great deal of room to spare.

Theorem 3.5. *Suppose that* $L = -\sum_j X_j^2$ *where the* X_j *are smooth real vector fields in some open set* V*, satisfying (2.2) for some* $w \in C^\infty$ *such that* $w(\xi) \to \infty$ *as* $|\xi| \to \infty$*. Suppose that the set of all points of* V *at which* $\{X_j\}$ *fails to satisfy the*

---

[8] M is a submanifold of $\partial\Omega$ of real codimension one.

[9] A C^ω analogue of Theorem 2.3 appears to be implicit in work of Grigis and Sjöstrand **[16]**, although we have not yet verified this in detail.

bracket hypothesis is totally disconnected. Then L is hypoelliptic in V. Moreover for any $s \in \mathbb{R}$, for any $u \in \mathcal{D}'(V)$, the H^s singular support of u is contained in the H^s singular support of Lu.

If in addition the set of points of S^*V at which the microlocal bracket hypothesis fails to be satisfied is totally disconnected, then the H^s wave front set of u is contained in the H^s wave front set of Lu.

This implies a more precise formulation of Proposition 1.4.

Corollary 3.6. *Suppose that $L = -\sum_j X_j^2$ where the X_j are smooth real vector fields in some open set V. Suppose that at each point of V, at least one X_j is nonzero. Suppose that $\{X_j\}$ satisfies the bracket hypothesis at all but finitely many points of V. Then for any $s \in \mathbb{R}$, for any $u \in \mathcal{D}'(V)$, the H^s singular support of u is contained in the H^s singular support of Lu.*

If in addition the set of points of S^*V at which the microlocal bracket hypothesis fails to be satisfied is finite, then the H^s wave front set of u is contained in the H^s wave front set of Lu.

Our final theorem is one of various possible parabolic analogues. Let $(x, t) \in \mathbb{R}^d \times \mathbb{R}^1$ be coordinates in $\mathbb{R}^{d+1}$.

Theorem 3.7. *Let $\{X_j\}$ be a collection of vector fields in $\mathbb{R}^d$ satisfying the hypotheses of Theorem 2.1. Then $\partial_t - \sum_j X_j^2$ is hypoelliptic in $\mathbb{R}^{d+1}$.*

Throughout the discussion we assume L to satisfy a compactness inequality (2.2) with $w \to \infty$. It is likely that a variant of Theorem 2.3 may be formulated and proved, in which only an inequality $\|u\|_{H^{-m}} \leq C\|Lu\|_{H^0}$ is assumed; a correspondingly stronger hypothesis depending on m must be imposed on ψ and the commutator operator to which it gives rise. Although examples are certainly known of hypoelliptic operators that satisfy only such weaker inequalities, we have not investigated the usefulness of such a generalization in the analysis of concrete examples.

It is not clear to this author to what extent the hypothesis $\psi \in S^0_{1,0}$ in Theorem 2.3 is natural. Perhaps a variant in which ψ is permitted to belong to a less restricted symbol class might also be useful in this context. Such variants appear for instance in **[19]**, **[25]**, **[26]**.

For remarks and speculation concerning conditions on symbols characterizing hypoellipticity, see **[34]** and **[10]**.

4. Proofs of Theorems 2.1 and 2.3

By a pseudodifferential operator we will always mean an operator of the form

$$\operatorname{Op}(a)f = \int_{\mathbb{R}^d} e^{ix\xi}\, a(x, \xi) \hat{f}(\xi)\, d\xi,$$

associated to the symbol a via the Kohn–Nirenberg calculus. These operators act on compactly supported distributions. We do not always assume them to be properly supported in the sense that $a(x,\xi) \equiv 0$ for x outside a compact subset of $\mathbb{R}^d$. However, pseudolocality ensures that if $U_1 \Subset U_2 \subset \mathbb{R}^d$ are open, and if A, B are pseudodifferential operators whose symbols each belong to one of the standard symbol classes $S^m_{\rho,\delta}$ with $\rho > 0$, and if $\eta \in C_0^\infty(\mathbb{R}^d)$ is $\equiv 1$ in a neighborhood of the closure of U_2, then as an operator from $\mathcal{E}'(U_1)$ to $\mathcal{D}'(U_2)$, $A \circ \eta \circ B$ is independent of η, modulo an operator mapping $\mathcal{E}'(U_1)$ to $C^\infty(U_2)$. Consequently when analyzing hypoellipticity in an open set V we will sometimes write $A \circ B$ to mean $A \circ \eta \circ B$, where η is a cutoff function that is identically equal to one in a neighborhood of $\overline{V}$.

Denote by $\sigma(P)$ a symbol of a pseudodifferential operator P. We say that an operator is smoothing of order M in the Sobolev scale if it maps H^s to H^{s-M} for all $s \in \mathbb{R}$. A natural setting[10] for much of our reasoning will be the classes $S^{m,n} \subset S^{m+}_{1,0}$, which by definition consist of all $a \in C^\infty(\mathbb{R}^d \times \mathbb{R}^d)$ satisfying

$$|\partial_x^\alpha \partial_\xi^\beta a(x,\xi)| \le C_{\alpha,\beta} \langle \xi \rangle^{m-|\beta|} [\log\langle \xi \rangle]^{n+|\alpha|+|\beta|}$$

for all $\alpha, \beta, (x,\xi)$. Certain manipulations performed below without comment are justified by well known basic properties of the operators associated to symbols in the latter class, such as the symbolic calculus and pseudolocality **[32]**.

We begin with the proof of Theorem 2.1, then will later indicate how it should be modified to derive the more general Theorem 2.3. Let $u \in \mathcal{D}'(V)$ and $s \in \mathbb{R}$ be given. Suppose the H^s wave front set of Lu to be disjoint from some conic open neighborhood Γ_0 of $(x_0, \xi_0) \in T^*V$. Without loss of generality we may assume $u \in \mathcal{E}'(V)$. Fix K such that $u \in H^{-K}$. Given any $s \in \mathbb{R}$, our aim is to show that $(x_0,\xi_0) \notin WF_{H^s}(u)$ by constructing a pseudodifferential operator Λ which is elliptic of order s in a conic neighborhood of (x_0,ξ_0), and for which it can be shown that $\Lambda u \in H^0(\mathbb{R}^d)$.

To do this fix a small conic open neighborhood Γ_1 of (x_0,ξ_0) whose intersection with the unit cosphere bundle is a compact subset of Γ_0. Fix an auxiliary function $\phi(x,\xi) \in C^\infty(\mathbb{R}^d \times \mathbb{R}^d \setminus \{0\})$ that is homogeneous of degree zero with respect to ξ, is everywhere nonnegative, vanishes identically in a small conic neighborhood of (x_0,ξ_0), and is strictly positive on the complement of Γ_1.

Define a symbol of nonconstant order, depending on parameters s, N_0, by

$$\lambda(x,\xi) = |\xi|^s e^{-N_0 \log|\xi| \phi(x,\xi)} \tag{4.1}$$

where $|\xi| \ge e$, and define λ for $|\xi| < e$ so as to be C^∞ and nowhere vanishing. The nonnegativity of ϕ implies that $\lambda \in S^{s+}_{1,0}$. Moreover, $\lambda \in S^{s,0}$.

With s fixed, there exists $\delta > 0$ such that for each N_0, $\lambda \in S^{-\delta N_0 +}_{1,0}$ on the closure of the complement of Γ_1. Choose N_0 so that $-\delta N_0 < -K$. Then $\Lambda u \in H^{-K+\delta N_0} \subset H^0$ microlocally on the complement of Γ_1.

[10] Instead of the symbol class $S^{m,n}$ and Kohn–Nirenberg quantization one could employ the Weyl calculus of Hörmander **[18]**, associated to the metric $g = \log^2\langle\xi\rangle\, dx^2 + \log^2\langle\xi\rangle \langle\xi\rangle^{-2}\, d\xi^2$.

Now L acts on sections of some bundle, and each X_j maps sections of that bundle to sections of some bundle W_j, whose rank may well depend on j. Restricting the analysis to a small open subset of $\mathbb{R}^d$, we may assume all these bundles to be trivial, and we fix trivializations of them. Define $\Lambda = \mathrm{Op}(\lambda \cdot I)$ where I denotes the identity matrix. Likewise define $\Lambda_j = \mathrm{Op}(\lambda \cdot I_j)$, acting on sections of W_j, where I_j is an identity matrix whose dimension equals the rank of W_j.

Fix cutoff functions $\eta_k \in C_0^\infty(\mathbb{R}^d)$ such that $\eta_2 \equiv 1$ in a neighborhood of the support of u, $\eta_1 \equiv 1$ in a neighborhood of the support of η_2, and η_1 is supported in V.

Recall that if a, b are symbols in some classes $S^m_{\rho,\delta}$ and $S^n_{\rho,\delta}$, if b is properly supported, and if $\rho > \delta$, then $\mathrm{Op}(a) \circ \mathrm{Op}(b)$ has a symbol $a \odot b$ with an asymptotic expansion

$$a \odot b(x,\xi) \sim \sum_\alpha c_\alpha \partial_\xi^\alpha a(x,\xi) \partial_x^\alpha b(x,\xi) \tag{4.2}$$

where $c_\alpha = (\alpha!)^{-1}(-i)^\alpha$. The notation $\sim$ indicates convergence in the usual asymptotic sense: for any positive integer N, the difference between $\mathrm{Op}(a) \circ \mathrm{Op}(b)$ and an operator associated to the symbol $\sum_{|\alpha|<N} c_\alpha \partial_\xi^\alpha a(x,\xi) \partial_x^\alpha b(x,\xi)$ is smoothing of order $m + n - N(\rho - \delta)$ in the scale of Sobolev spaces.

Lemma 4.1. *There exists an operator, denoted Λ^{-1}, belonging to $S^{m+}_{1,0}$ for some $m = m(s)$, such that $\Lambda \circ \Lambda^{-1} - I$ is smoothing of infinite order in the Sobolev scale. Moreover such an operator may be constructed with a symbol of the form $(1+f)\lambda^{-1} \cdot I$, where $f \in S^{-1,2}$. Likewise there exist operators Λ_j^{-1} of the same form $(1+f_j)\lambda^{-1} \cdot I_j$, such that each $\Lambda_j \circ \Lambda_j^{-1} - I_j$ is smoothing of infinite order.*

Proof. Write $f \sim \sum_{k\geq 1} f_k$. Solve the equation $\lambda \odot ((1+f)\lambda^{-1}) \sim 1$, using (4.2) and the usual iterative procedure. One obtains $f_1 \in S^{-1,2}$, and by induction, each $f_k \in S^{-k+}_{1,0}$. Choose Λ to be an operator whose full symbol has asymptotic expansion $\sum_{k\geq 1} f_k$; such operators exist **[31]**, **[32]**. The same method of construction yields Λ_j. □

Lemma 4.2. *Suppose that L takes the form* (2.1). *Then there exists a pseudodifferential operator G of the form*

$$G = \sum_j B_j \circ X_j + \sum_j X_j^* \circ \tilde{B}_j + B_0 \tag{4.3}$$

where $B_0 \in \mathrm{Op}(S^{0,2})$ and $B_j, \tilde{B}_j \in \mathrm{Op}(S^{0,1})$ for each $j \geq 1$, such that

$$(L+G)\eta_1 \Lambda \eta_2 = \eta_1 \Lambda L \eta_2 + R \tag{4.4}$$

for some R belonging to $\mathrm{Op}(S^{-M+}_{1,0})$ for every $M < \infty$.

Proof of Lemma 4.2. Throughout the argument, R denotes an operator belonging to $S^{-M+}_{1,0}$ for all M, which may change from one line to the next. In constructing the symbol of G we work formally, ignoring the cutoff functions η_j; this is permissible by pseudolocality, because $\eta_2\eta_1 \equiv \eta_2$. The desired equation $(L+G)\Lambda = \Lambda L + R$ is then equivalent to

$$G = \Lambda L \Lambda^{-1} - L + R\Lambda^{-1} = \sum_j \left[\Lambda X_j^* X_j \Lambda^{-1} - X_j^* X_j\right] + R\Lambda^{-1}$$

$$= \sum_j \left[(\Lambda X_j^* \Lambda_j^{-1})(\Lambda_j X_j \Lambda^{-1}) - X_j^* X_j\right] + R\Lambda^{-1}.$$

Consider $\Lambda_j X_j \Lambda^{-1} = X_j + (\Lambda_j X_j - X_j\Lambda)\Lambda^{-1} + R$. The symbol of $\Lambda_j X_j - X_j \Lambda$ divided by λ equals the Poisson bracket $\{\log\lambda,\ \sigma(X_j)\}$ plus an element of $S^{-1,2}$, as follows from (4.2). Applying the composition formula (4.2) once more we conclude

$$\Lambda_j X_j \Lambda^{-1} = X_j + \mathrm{Op}(\{\log\lambda,\ \sigma(X_j)\}) \quad \text{modulo } \mathrm{Op}(S_{1,0}^{-1+}). \tag{4.5}$$

A corresponding assertion holds for $\Lambda X_j^* \Lambda_j^{-1}$, with $\sigma(X_j^*)$ substituted for $\sigma(X_j)$ in the Poisson bracket. Since $\{\log\lambda, \sigma(X_j)\}$ and $\{\log\lambda,\ \sigma(X_j^*)\}$ belong to $S^{0,1}$, inserting these equations into the identity derived for G in the preceding paragraph completes the proof. □

Lemma 4.3. *Let G be a pseudodifferential operator of the form* (4.3). *Then for any fixed relatively compact subset $U \subset V$, any $\delta > 0$ and any $f \in C_0^\infty$ supported in U,*

$$|\langle Gf,\ f\rangle| \le C_\delta \int \log^2\langle\xi\rangle\, |\hat{f}(\xi)|^2\, d\xi + \delta \sum_j \|X_j f\|_{L^2}^2. \tag{4.6}$$

This lemma follows from Gårding's inequality (and its proof, adapted to $\mathrm{Op}(S_{1,0}^{1+})$). One uses the nonnegativity of $C\log^2\langle\xi\rangle$ minus the symbol of B_0, provided that C is chosen to be sufficiently large, along with the inclusions $B_j,\ B_j^* \in S^{0,1}$. The detailed verification is left to the reader.

Lemma 4.4. *Let L take the form* (2.1) *and satisfy* (2.4). *Let $s, M \in \mathbb{R}$ be fixed. If N_0 is chosen to be sufficiently large in the definition of Λ, then for any fixed relatively compact subset $U \Subset V$ and any $u \in C^{s+3}(U)$,*

$$\|\eta_1 \Lambda u\|_{L^2} \le C\|\eta_1 \Lambda L u\|_{L^2} + C\|u\|_{H^{-M}}. \tag{4.7}$$

All norms without subscripts in the following argument are $L^2 = H^0$ norms.

Proof. We have $(L+G)\eta_1\Lambda\eta_2 u = \eta_1\Lambda L\eta_2 u$, and $\eta_2 u \equiv u$. Thus setting $v = \eta_1 \Lambda u \in C^2$,

$$\big\langle (L+G)v,\ v\big\rangle = \sum_j \|X_j v\|^2 + \|v\|^2\big) + O\big(\|v\| \cdot \|Gv\|\big).$$

Invoking Lemma 4.3 we find that

$$\sum_j \|X_j v\|^2 \le \|\eta_1 \Lambda L u\|^2 + \|Ru\|^2 + C\|v\|^2 + C\|Gv\|^2$$

$$\le \|\eta_1 \Lambda L u\|^2 + \|Ru\|^2 + C\|v\|^2 + C_\delta \int \log^2\langle\xi\rangle\, |\hat{v}(\xi)|^2\, d\xi + \delta \sum_j \|X_j v\|^2.$$

The error term R maps H^{-M} to L^2, and $\|v\|^2$ is majorized by $\int \log^2\langle\xi\rangle |\hat{v}|^2 d\xi$. Choosing $\delta < 1$, the last term on the right hand side may be absorbed into the left, yielding

$$\sum_j \|X_j v\|^2 \le C \int \log^2\langle\xi\rangle\, |\hat{v}(\xi)|^2\, d\xi + \|\eta_1 \Lambda L u\|^2 + C\|u\|^2_{H^{-M}}.$$

We finally invoke the hypothesis (2.4) in the form

$$\sum_j \|X_j v\|^2 \ge A \int \log^2\langle\xi\rangle\, |\hat{v}(\xi)|^2\, d\xi - C_A\|v\|^2$$

for arbitrarily large A to deduce that

$$\int \log^2\langle\xi\rangle\, |\hat{v}(\xi)|^2\, d\xi \le C\|\eta_1 \Lambda L u\|^2 + C\|u\|^2_{H^{-M}} + C\|v\|^2$$

for some constant C. $\|v\|$ may be majorized by an arbitrarily small constant times the left hand side plus a large constant times the H^{-M-s} norm of v. $\eta_1 \Lambda \eta_2$ maps H^{-M} boundedly to H^{-M-s}, so $\|v\|_{H^{-M-s}} \le C\|u\|_{H^{-M}}$. Thus

$$\|v\|^2 \le \int \log^2\langle\xi\rangle |\hat{u}(\xi)|^2\, d\xi \le C\|\eta_1 \Lambda L u\|^2 + C\|u\|^2_{H^{-M}}. \qquad \square$$

Proof of Theorem 2.1. It remains to remove the smoothness assumption on u and to convert the *a priori* estimate to the desired conclusion $\Lambda u \in H^0$. To accomplish this fix an auxiliary function $r \in C^\infty(\mathbb{R}^d)$ that is strictly positive, and satisfies $r(\xi) \equiv |\xi|^{-1}$ for all $|\xi| \ge 2$, and $r(\xi) \equiv 1$ for all $|\xi| \le 1$. Fix a large exponent q. For all small $\varepsilon > 0$ define a mollified symbol

$$\lambda_\varepsilon(x, \xi) = r^q(\varepsilon\xi) \cdot \lambda(x, \xi) \tag{4.8}$$

where λ is as defined in (4.1). and let $\Lambda_\varepsilon = \mathrm{Op}(\lambda_\varepsilon)$. The symbols $r_\varepsilon = r^q(\varepsilon\xi)$ satisfy

$$|\partial_\xi^\alpha r_\varepsilon / r_\varepsilon| \le C_{\alpha,q} |\xi|^{-|\alpha|}, \quad \text{uniformly in } \varepsilon \text{ and } \xi \in \mathbb{R}^d. \tag{4.9}$$

If q is chosen to be sufficiently large relative to the order of the distribution u, then $\Lambda_\varepsilon u \in C^2$ for all $\varepsilon > 0$, and because λ is elliptic of order s in a conic neighborhood of (x_0, ξ_0), it suffices to show that the L^2 norm of $\eta_1 \Lambda_\varepsilon u$ remains uniformly bounded as $\varepsilon \to 0$. But Lemma 4.4 fails to apply directly, because u is merely known to be a distribution, not a function in C^{s+3} as hypothesized.

The parameter N_0 in (4.8) may be chosen sufficiently large that $\eta_1 \Lambda L u \in L^2$, because ϕ is strictly positive in a conic neighborhood of the H^s wave front set of u, and hence Λ is regularizing there of order at least $s - \delta N_0$ for some constant $\delta > 0$. The L^2 norm of $\eta_1 \Lambda_\varepsilon L u$ is bounded uniformly in ε and tends to the L^2 norm of $\eta_1 \Lambda L u$.

As in the proof of Lemma 4.4 we have for each $\varepsilon > 0$ an operator G_ε and an identity $(L + G_\varepsilon)\eta_1 \Lambda_\varepsilon u = \eta_1 \Lambda_\varepsilon L u + R u$ with both sides of the equation in C^2 for each $\varepsilon > 0$. Moreover (4.9) ensures that the proof of Lemma 4.2 carries through for each $\varepsilon > 0$ with Λ replaced by Λ_ε, so that G_ε takes the form (4.3), with each pseudodifferential coefficient B_j in the class indicated in Lemma 4.2, and with all bounds uniform in ε. All functions have sufficient differentiability for the proof of Lemma 4.4 and this identity to yield

$$\|\eta_1 \Lambda_\varepsilon u\|_{L^2} \le C \|\eta_1 \Lambda_\varepsilon L u\|_{L^2} + C \|u\|_{H^{-M}},$$

uniformly as $\varepsilon \to 0$. Thus the L^2 norm of $\eta_1 \Lambda_\varepsilon u$ remains bounded as $\varepsilon \to 0$. □

The proof of Lemma 4.2 yields the following more precise conclusion, which will be used to prove Theorem 2.3.

Lemma 4.5. *Let Λ be any pseudodifferential operator whose symbol λ takes the general form $\lambda(x,\xi) = |\xi|^s \exp\big(-N \log|\xi| \psi(x,\xi)\big)$ for large $|\xi|$ where $\psi \in S^0_{1,0}$. Suppose that L takes the form* (2.1). *Define $b_j = \mathrm{Op}(\{\log\lambda, \sigma(X_j^*)\})$ and $\tilde{b}_j = \mathrm{Op}(\{\log\lambda, \sigma(X_j)\})$. Then there exists a pseudodifferential operator G of the form* (4.3) *satisfying* (4.4), *with*

$$\begin{aligned} &B_j = b_j + c_j \text{ and } \tilde{B}_j = \tilde{b}_j + \tilde{c}_j \quad \text{for every } j \ge 1 \\ &B_0 = \sum_j \big(b_j \circ \tilde{b}_j + A_j \tilde{b}_j + \tilde{A}_j b_j\big) \quad \text{modulo } \mathrm{Op}(S^{-1,2}) \end{aligned} \tag{4.10}$$

where each $c_j, \tilde{c}_j \in \mathrm{Op}(S^{-1,1})$, and where $A_j, \tilde{A}_j \in S^0_{1,0}$ are the coefficients in (2.1).

Lemma 4.6. *Suppose that L, ψ, p satisfy the hypotheses of Theorem* 2.3, *and that*

$$\lambda(x,\xi) = |\xi|^s \exp(-N \log\langle\xi\rangle \psi(x,\xi)) \quad \text{for large } |\xi|. \tag{4.11}$$

Then for any $N \ge 0$ and for any fixed relatively compact subset $U \subset V$, any $\delta > 0$ and any $f \in C^{s+3}$ supported in U, the operator G constructed in Lemma 4.2 *satisfies*

$$|\langle Gf, f\rangle| \le \delta \sum_j \|X_j f\|^2 + C_\delta \|f\|^2 + C_\delta \|\mathrm{Op}(p) f\|^2_{H^1}. \tag{4.12}$$

This follows directly from (4.10) and the hypothesis (2.7).

Lemma 4.7. *Let L, ψ, p satisfy the hypotheses of Theorem* 2.3, *and let λ be defined by* (4.11). *Let s, M be fixed. If N is chosen to be sufficiently large in the definition* *of* λ *then for any fixed relatively compact subset* $U \subset V$ *and any* $u \in C_0^\infty(U)$,

$$\|\eta_1 \Lambda u\|_{L^2} \le C\|\eta_1 \Lambda L u\|_{L^2} + C\|u\|_{H^{-M}}.$$

This follows from the same reasoning as in the proof of Lemma 4.4, using the hypotheses on ψ and Lemma 4.6. The smoothness assumption can be removed, and the a priori estimate converted to the conclusion $\eta_1 \Lambda u \in L^2$, as above. The term involving the H^1 norm of $\mathrm{Op}(p)\Lambda u$ may be absorbed into $\|u\|_{-M}$, because Λ may be made to be regularizing of arbitrarily high order in a conic neighborhood of the support of the symbol p, by choosing N to be sufficiently large. This completes the proof of Theorem 2.3. Theorem 2.4 is proved in the same way.

5. Analysis of examples

The purpose of this section is to analyze the three classes of examples L_j discussed in the introduction. It will be convenient to assume that $a(x) \to \infty$ as $|x| \to \infty$; this has no effect on the question of hypoellipticity. Fix a and define $\mathcal{L} = \mathcal{L}_\tau = -\partial_x^2 + \tau^2 a^2(x)$. Define $\lambda_0 = \lambda_0(\tau)$ to the infimum of $\langle \mathcal{L} f, f\rangle^{1/2}/\|f\|_{L^2}$ over all $0 \ne f \in C_0^2(\mathbb{R})$. The notation $A \sim B$ will mean that A/B is bounded above and below by positive constants, independent of τ as $\tau \to +\infty$.

Lemma 5.1.

$$\lambda_0 \sim \min_{0<y<\infty} \big(y^{-1} + \tau a(y)\big). \tag{5.1}$$

Proof. To obtain the upper bound for λ_0 it suffices to fix a test function $\varphi \in C_0^1(\mathbb{R})$ supported in $\{|x| < 1\}$, and to consider $\varphi(x/y)$ for arbitrary y. The conclusion is that $\lambda_0 \le C(y^{-1} + \max_{|x|\le y} \tau a(x))$, which is the required upper bound because of the monotonicity of a.

For a lower bound, consider any $y > 0$. For any $\varphi \in C_0^2$, $\langle \mathcal{L}\varphi, \varphi\rangle = \|\partial_x \varphi\|^2 + \int \tau^2 a^2(x)\,\varphi^2(x)\,dx$. Now

$$\int_{|x|\ge y} \tau^2 a^2 \varphi^2\,dx \ge \tau^2[\min_{|x|\ge y} a^2(x)] \int_{|x|\ge y} \varphi^2\,dx.$$

Also

$$\int_{|x|<y} \varphi^2\,dx \le C\int_{y<|x|<2y} \varphi^2\,dx + Cy^2\|\partial_x\varphi\|^2.$$

Thus

$$\|\varphi\|^2 \le C\big(\tau^{-2}[\min_{|x|\ge y} a(x)]^{-2} + y^2\big)\langle \mathcal{L}\varphi, \varphi\rangle \quad \text{for any } y > 0. \tag{5.2}$$

We now invoke the monotonicity of a to choose the unique y satisfying $y^{-1} = \tau a(y)$. Then y^{-1} is comparable to the minimum in (5.1), since the functions x^{-1} and $\tau a(x)$ are respectively nonincreasing and nondecreasing on $\mathbb{R}^+$. Thus $(y^2 + \tau^{-2} a(y)^{-2}) \le C(y^{-1} + \tau a(y))^{-2}$, as desired. □

Lemma 5.2. *Assume that a is even and nonnegative, vanishes only where $x = 0$, is nondecreasing on $[0, \infty)$ and that $a(x) \to \infty$ as $x \to +\infty$. Then*

$$\lim_{\tau\to\infty} \frac{\lambda_0(\tau)}{\log \tau} = \infty \iff \lim_{x\to 0} x \log a(x) = 0.$$

Proof. Given any large $\tau \in \mathbb{R}^+$, define $y \in \mathbb{R}^+$ by $y^{-1} = \tau a(y)$. Then $\lambda_0(\tau) \sim y^{-1}$, as observed in the proof of Lemma 5.1. Thus

$$\log \tau / \lambda_0(\tau) \sim y \log \tau = y |\log[y a(y)]| \sim y \log a(y).$$

Thus if $x \log a(x) \to 0$ as $x \to 0$, then $\lambda_0(\tau)/\log \tau \to \infty$ as $\tau \to \infty$.

Conversely if a small $x > 0$ is given, define $\tau = x^{-1} a(x)^{-1}$. Then $\tau \to \infty$ as $x \to 0$, and again $\lambda_0(\tau) \sim x^{-1}$, so

$$\lambda_0(\tau)/\log \tau \sim x^{-1}/\log(x^{-1} a(x)^{-1}) \sim |x \log a(x)|^{-1}.$$

The converse implication follows directly. □

In the next lemma we omit the hypothesis that a is nondecreasing.

Lemma 5.3. *If $x \log a(x) \to 0$ as $x \to 0$ and $a(x) \to \infty$ as $|x| \to \infty$, then $\lambda_0(\tau)/\log \tau \to \infty$ as $\tau \to \infty$.*

Proof. The proof of (5.2) did not utilize the hypothesis that a was nondecreasing. Given merely that $\lim_{x\to 0} x \log a(x) = 0$, then for each $\delta > 0$ there exists $c_\delta > 0$ such that $a(x) \ge c_\delta \exp(-\delta/2|x|)$ as $x \to 0$. When $y = \delta/\log \tau$, the factor on the right in (5.2) becomes

$$y + \tau^{-1}[\min_{|x|\ge y} a(x)]^{-1} \le \delta/\log \tau + C_\delta \tau^{-1} \exp(\delta \log \tau / 2\delta) \le C\delta/\log \tau$$

for all sufficiently large τ, with C' independent of δ, τ. Given any τ so large, applying (5.2) with these choices of δ and of y yields $\lambda_0(\tau) \ge c\delta^{-1} \log \tau$. □

Lemma 5.4. *For any coefficient $a \in C^\infty$, if L_2 is hypoelliptic in some neighborhood of 0, then $\lambda_0(\tau)/\log \tau \to \infty$ as $\tau \to +\infty$.*

Proof. The coefficient $a(x)$ may be assumed to tend to $+\infty$ as $|x| \to \infty$. Then each operator $\mathcal{L}_\tau$ is an essentially self-adjoint operator on $L^2(\mathbb{R})$, with a discrete spectrum tending to $+\infty$. Its lowest eigenvalue is $\lambda_0^2(\tau)$. There exists an associated eigenfunction f_τ that is strictly positive everywhere, is even, and assumes its maximum value at $x = 0$. Normalize it so that $f_\tau(0) = 1$. Define a solution to $LF_\tau \equiv 0$ by

$$F_\tau(x, y, t) = e^{i\tau t} e^{\lambda_0(\tau) y} f_\tau(x).$$

Suppose L_2 to be hypoelliptic in some small bounded neighborhood V of 0. Then by the Baire category theorem, for each positive integer k there must exist $C < \infty$ such that for every $F \in C^0(V)$ satisfying $L_2 F \equiv 0$ in V,

$$|\partial_t^k F(0)| \leq C\|F\|_{L^\infty(V)}. \tag{5.3}$$

Considering any large $\tau \in \mathbb{R}^+$ and taking $F = F_\tau$, the right hand side of (5.3) is bounded by $C\exp(\lambda_0(\tau))$ since $|f_\tau| \leq 1$. The left hand side equals $|\tau^k f_\tau(0)| = \tau^k$. Therefore for any k there must exist $C_k < \infty$ such that $\tau^k \leq C_k e^{\lambda_0(\tau)}$. Thus $\lambda_0(\tau)/\log\tau \geq k - \log C_k/\log\tau$, whence $\liminf_{\tau\to\infty}(\lambda_0(\tau)/\log\tau) \geq k$. □

This argument is parallel to that used by many authors to disprove analytic or Gevrey class hypoellipticity for various classes of examples.

Proof of Proposition 1.2. Combining Lemmas 5.2 and 5.4, we find that if L_2 is hypoelliptic then $x\log a(x)$ tends to 0 as $x \to 0$.

Let $(x, t; \xi, \tau)$ be coordinates in $T^*\mathbb{R}^2$. If $x\log a(x) \to 0$, then $\lambda_0(\tau)/\log\tau \to \infty$. To relate this information to $\iint \log^2\langle(\xi,\tau)\rangle|\hat{u}(\xi,\tau)|^2\,d\xi d\tau$, observe that L_2 is elliptic except where $\xi \neq 0$. Therefore by a microlocalization it suffices to majorize the integral over the region $|\tau| \geq |\xi|$. Replacing $\log\langle(\xi,\tau)\rangle$ by $\log\langle\tau\rangle$,

$$\iint \log^2\langle\tau\rangle|\hat{u}(\xi,\tau)|^2\,d\xi d\tau = c\iint \log^2\langle\tau\rangle|\tilde{u}(x,\tau)|^2\,dx d\tau$$

where $\tilde{u}$ denotes the partial Fourier transform with respect to the second coordinate. For each τ,

$$\log^2\langle\tau\rangle\int|\tilde{u}(x,\tau)|^2\,dx \leq \log^2\langle\tau\rangle\lambda_0(\tau)^{-2}\int|(X_1u(x,\tau))\tilde{}|^2 + |(X_2u(x,\tau))\tilde{}|^2\,dx.$$

Because $\log\tau/\lambda_0(\tau) \to 0$, this leads to the superlogarithmic inequality (2.4) by splitting the analysis into two cases $|\tau| \geq A$ and $|\tau| < A$, and choosing $A = A(\delta)$ to tend to ∞ sufficiently rapidly as $\delta \to 0$.

Lastly (2.4) implies that L_2 is hypoelliptic, by the general Theorem 2.1. □

The reasoning used above to estimate $\lambda_0(\tau)$ can be modified to prove that if $a \in C^\infty$ vanishes so rapidly at 0 that for any $A < \infty$ there exists C_A such that $|a(x)| \leq C_A\exp(-A|x|^{-1})$ as $x \to 0$, then $\lambda_0(\tau)/\log\tau \to \infty$, and consequently L_2 fails to be hypoelliptic.

Proof of Proposition 1.1. Write $L_1 = -\partial_{x_1}^2 - a^2(x_1)\partial_{x_2}^2$ in coordinates $x = (x_1, x_2)$, and let $\xi = (\xi_1, \xi_2)$ be dual coordinates. Set $X_1 = \partial_{x_1}$, $X_2 = a(x_1)\partial_{x_2}$. Let $R = \{(x,\xi) : x = 0,\ \xi_1 = 0,\ \xi_2 > 0\}$. The principal symbol of L_1 vanishes precisely on the symplectic manifold where $x_1 = \xi_1 = 0$. Since L_1 is invariant under translation with respect to x_2, and reflection about the origin, it suffices to prove that $Lu \in H^s(R) \Rightarrow u \in H^s(R)$.

In order to apply Theorem 2.3 we set $p \equiv 0$ and seek to construct ψ having favorable commutation properties. Given any number $\rho > 0$, there exists $\psi \in$

$C^\infty(T^*V)$ that is homogeneous of degree zero with respect to ξ, is $\equiv 1$ where $|(x_1, x_2, \xi_1/\xi_2)| \geq 3\rho$, is $\equiv 0$ where $|(x_1, x_2, \xi_1/\xi_2)| \leq \rho$, and depends only on x_2 where $|(x_1, \xi_1/\xi_2)| \leq 2\rho$. In order to apply Theorem 2.3, we must verify that $\mathrm{Op}[\log\langle\xi\rangle\{\psi, \sigma(X_i)\}]$ is controlled by X_1, X_2 in the sense (2.7). Microlocally where $(x_1, \xi_1) \neq 0$, $L = X_1^2 + X_2^2$ is elliptic, so we have better control than is needed. Therefore it suffices to work microlocally where $|(x_1, \xi_1/\xi_2)| < \rho$.

In this region $\{\psi, \sigma(X_1)\} \equiv 0$, while $\{\psi, \sigma(X_2)\}(x, \xi) \equiv ia(x_1)\partial\psi/\partial x_2$. Since $\partial\psi/\partial x_2 \in S^0_{1,0}$ and because of the microlocalization to $|\xi_1| \ll \xi_2$, it suffices to verify that the quantity $\|\mathrm{Op}[\log\langle\xi_2\rangle a(x_1)]u\|$ is majorized by $\delta\|X_2 u\| + C_\delta\|u\|$, for all u supported where $|x| < 3\rho$. This majorization follows from an application of the partial Fourier transform with respect to x_2, since $\log\langle\xi_2\rangle/\xi_2 \to 0$ as $\xi_2 \to +\infty$. □

Proof of Proposition 1.3. Let $x = (x_1, x_2, x_3)$, $\xi = (\xi_1, \xi_2, \xi_3)$ and $L_3 = -\partial_{x_1}^2 - b^2(x_1)\partial_{x_2}^2 - a^2(x_1)\partial_{x_3}^2 = -X_1^2 - X_2^2 - X_3^2$. Now the principal symbol of L_3 vanishes only where $x_1 = \xi_1 = 0$. As in the proof of Proposition 1.1, it suffices to work near where $x_1 = x_2 = \xi_1 = 0$. Again let $p \equiv 0$. Given any small number $\rho > 0$ we may construct ψ so as to have all properties required of it in Theorem 2.3, and in addition to be independent of x_1, ξ_1 where $|(x_1, \xi_1/|\xi|)| < \rho$. Microlocalizing to a small conic neighborhood of the nonelliptic region for L_3 as in the proof of Proposition 1.1, we have $\{\psi, \sigma(X_1)\} \equiv 0$, $\{\psi, \sigma(X_2)\} \equiv ib(x_1)\partial\psi/\partial x_2$, and $\{\psi, \sigma(X_3)\} \equiv ia(x_1)\partial\psi/\partial x_3$. Because $b \geq a$, the analysis is straightforward where $|\xi_2| \geq |\xi_3|$, using the partial Fourier transform with respect to (x_2, x_3).

Where $|\xi_3| \geq |\xi_2|$, we must majorize in terms of $\partial\tilde{u}/\partial x_1$ and $a(x_1)\xi_3\tilde{u}(x_1, \xi_2, \xi_3)$ the $L^2(dx_1)$ norm of the partial Fourier transform $\log\langle\xi_3\rangle b(x_1)\tilde{u}(x_1, \xi_2, \xi_3)$; the contribution of $X_2 u$ is of no use since ξ_2 could vanish.

It now suffices to show that under the hypothesis $b(x)x\log a(x) \to 0$, for all $\tau \in \mathbb{R}^+$

$$\log\tau\|b\varphi\| \leq \varepsilon(\tau)\|\varphi'\| + \varepsilon(\tau)\|a\tau\varphi\| \quad \text{for all } \varphi \in C_0^1(\mathbb{R}^1), \tag{5.4}$$

where $\varepsilon(\tau) \to 0$ as $\tau \to +\infty$ and each norm is that of $L^2(\mathbb{R}^1)$.

For the remainder of the discussion, x denotes an element of $\mathbb{R}^1$. In the course of the proof of Lemma 5.1 it was shown that for any smooth even coefficient a vanishing only at $x = 0$ and nondecreasing on $\mathbb{R}^+$,

$$\|\partial_x u\|^2 + \|\tau a \cdot u\|^2 \geq c\int \tau^2 \max(a^2(x), a^2(r))|u^2(x)|\,dx \quad \text{for all } u \in C_0^1(\mathbb{R}),$$

where $r = r(\tau)$ is chosen so that $r^{-1} = \tau a(r)$. Therefore (5.4) would be a consequence of the majorization

$$b(x)\log\tau \leq \varepsilon(\tau)\tau\max(a(x), a(r(\tau))) \tag{5.5}$$

uniformly in x, where $\varepsilon(\tau) \to 0$ as $\tau \to \infty$.

The definition of r implies that $\log\tau = \log r^{-1} + \log a^{-1}(r) \sim \log a^{-1}(r)$ and $\tau a(r) = r^{-1}$, so that when $x = r$, $[b(r)\log\tau]/[\tau a(r)] \sim b(r)r|\log a(r)| = \varepsilon(\tau)$, which indeed tends to zero as $\tau \to \infty$ by hypothesis. Since b is nondecreasing on $\mathbb{R}^+$, this implies that $b(x)\log\tau \le \varepsilon(\tau)\tau a(r)$ for all $|x| \le r$. Write $\delta(x) = b(x)x|\log a(x)|$. Then for $x > r$,

$$\frac{b(x)\log\tau}{\tau a(x)} \le \frac{\delta(x)\log\tau}{x|\log a(x)|\tau a(x)} \sim \delta(x) \cdot \frac{ra(r)|\log a(r)|}{xa(x)|\log a(x)|} \le C\delta(x)\frac{r}{x}.$$

If $x \ge r^{1/2}$ this is bounded by $Cr^{1/2}$, which tends to zero as $\tau \to \infty$. If $x \le r^{1/2}$ it is bounded by $\min_{0<t<r^{1/2}} \delta(t)$, which also tends to zero as $\tau \to \infty$. Thus (5.5) is indeed valid.

That hypoellipticity of L_3 implies (1.2) when b vanishes to finite order follows from reasoning similar to that underlying Lemmas 5.1 and 5.2. The details are left to the reader. □

The above analysis shows that the nonhypoelliptic example $L_2 = -\partial_x^2 - a^2(x)\partial_t^2 - \partial_y^2$, with coefficient $a(x) = \exp(-1/|x|)$, does satisfy the inequality (2.2) with the borderline weight $w(\xi) = \log\langle\xi\rangle$.

6. Proofs of Theorem 3.5 and Lemma 3.1

Proof of Lemma 3.1. Since the bracket hypothesis is satisfied on the complement of M, it is no loss of generality to assume U to be a small neighborhood of a point $x_0 \in M$, and to assume that $\beta(x) \ne 0$ for all $x \in \overline{U}\backslash M$.

Let $A = (I - \Delta)^{1/2}$. Fix U and consider any $u \in C_0^\infty(U)$. It is known **[32]** that for any k and any $Z \in \mathfrak{g}_k$,

$$\|ZA^{-1+2^{-k}}u\|^2 \le C\sum \|X_j u\|^2 + C\|u\|^2. \tag{6.1}$$

This is a key step in one proof of hypoellipticity of sums of squares operators satisfying Hörmander's bracket hypothesis **[32]**, but its proof does not require that hypothesis.

Consider any smooth vector field Y. By Cramer's rule we may express $Y = \sum_{i,\alpha} c_{i,\alpha} Z_{i,\alpha}$ where the coefficients c are smooth on $V\backslash M$ and are $O(\beta^{-1/2})$ as $x \to M$. Therefore $\beta(x)\nabla_x$ may be expressed as $\sum_{i,\alpha} \tilde{c}_{i,\alpha} Z_{i,\alpha}$ where the vector valued coefficients $\tilde{c}(x)$ are uniformly bounded on $\overline{U}$. Thus there exists $\delta > 0$ such that

$$\|\beta\nabla_x A^{-1+\delta}u\|^2 \le C\sum \|X_j u\|^2 + C\|u\|^2.$$

Since $\beta \in C^\infty$, this implies that

$$\|\beta A^\delta u\|^2 \le C\sum \|X_j u\|^2 + C\|u\|^2. \tag{6.2}$$

Fix coordinates $(s, y) \in \mathbb{R} \times \mathbb{R}^{d-1}$ in which $M = \{s = 0\}$, and in which one of the vector fields X_j equals ∂_s. In order to deduce the superlogarithmic gain estimate

(3.3), it now suffices[11] to show that under the hypothesis (3.2),

$$\int \log^2\langle\xi\rangle\, |\hat{u}(\xi)|^2\, d\xi \le \varepsilon\big(\|\partial_s u\|^2 + \|\beta A^\delta u\|^2\big) + C_\varepsilon \|u\|^2 \tag{6.3}$$

for all ε, u.

Fix an auxiliary function $h \in C_0^\infty(\mathbb{R}^+)$ that is supported on $[1/2, 4]$ and is identically equal to 1 on $[1, 2]$. Fix a second such function $\tilde{h} \in C_0^\infty(\mathbb{R}^+)$ that is identically equal to one on the support of h. Fix $\eta \in C_0^\infty(\mathbb{R}^n)$ supported in a small neighborhood of the closure of U, and $\equiv 1$ in a smaller neighborhood of its closure. Let P_t, $\tilde{P}_t$ be the pseudodifferential operators with symbols $\eta(x)h(|\xi|/t)$, $\eta(x)\tilde{h}(|\xi|/t)$, respectively. We seek a bound for $\log(t)\|P_t u\|$, for $t \ge e$.

The hypothesis (3.2) plus Lemma 5.3 with $\tau = t^\delta$ give

$$\log^2(t) \int |g(s)|^2\, ds \le \varepsilon(t) \int |\partial_s g|^2(s)\, ds + \varepsilon(t) t^{2\delta} \int \beta^2(s, y)|g|^2(s)\, ds$$

for all g supported in a sufficiently small neighborhood of $s = 0$, where $\varepsilon(t) \to 0$ as $t \to \infty$ and ε is a nonincreasing function of t. This inequality is uniform in y. Setting $g(s) = P_t u(s, y)$ and integrating both sides of the preceding inequality with respect to y yields

$$\log^2(t)\|P_t u\|^2 \le \varepsilon(t)\|\partial_s P_t u\|^2 + \varepsilon(t) t^{2\delta}\|\beta P_t u\|^2 + C_N t^{-N}\|u\|^2$$

for all N; this last term arises because $P_t u$ will not have compact support and hence must be truncated. Similar terms in estimates below arise in the same way. By pseudodifferential calculus,

$$\|\partial_s P_t u\|^2 + t^{2\delta}\|\beta P_t u\|^2 \le \|\tilde{P}_t \partial_s u\|^2 + C\|\tilde{P}_t \beta A^\delta u\|^2 + C_N t^{-N}\|u\|^2.$$

Therefore choosing any large parameter T,

$$\begin{aligned}
\int \log^2\langle\xi\rangle\, |\hat{u}(\xi)|^2\, d\xi &\sim \int_e^\infty \log^2(t)\|P_t u\|^2 \tfrac{dt}{t} + \|u\|^2 \\
&\le \int_e^\infty \varepsilon(t)\|\tilde{P}_t \partial_s u\|^2 \tfrac{dt}{t} + \int_e^\infty \varepsilon(t)\|\tilde{P}_t \beta A^\delta u\|^2 \tfrac{dt}{t} + C\|u\|^2 \\
&\le \varepsilon(T) \int_T^\infty \big(\|\tilde{P}_t \partial_s u\|^2 + \|\tilde{P}_t \beta A^\delta u\|^2\big) \tfrac{dt}{t} + C_T\|u\|^2 \\
&\le C\varepsilon(T)\big(\|\partial_s u\|^2 + \|\beta A^\delta u\|^2\big) + C_T\|u\|^2 \\
&\le C\varepsilon(T) \sum_j \|X_j u\|^2 + C_T\|u\|^2.
\end{aligned}$$

Choosing T so that $C\varepsilon(T) \le \delta$ yields the desired bound $\delta \sum_j \|X_j u\|^2 + C_\delta\|u\|^2$. □

[11] The order of magnitude of the quantity $\int \log^2\langle\xi\rangle\, |\hat{u}(\xi)|^2\, d\xi$ is invariant under changes of coordinates. This is a consequence of diffeomorphism invariance of the classes $\mathrm{Op}(S^m_{\rho,\delta})$ for (ρ, δ) sufficiently close to $(1, 0)$; see Theorem 5.1, Chapter II of [31].

Coupling the next lemma with Theorem 3.5 directly implies Corollary 3.6. The symbol $\|\cdot\|$ with no subscript will always denote the L^2 norm.

Lemma 6.1. *Suppose that a finite collection $\{X_j\}$ of C^∞ real vector fields satisfies the bracket condition at every point of $V\backslash\{x_0\}$, and that $X_1(x_0) \neq 0$. Let $U \Subset V$ be relatively compact. Then for every $\delta > 0$, for all $u \in H^1$ supported in U,*

$$\|u\| \leq \delta \sum \|X_j u\| + C_\delta \|u\|_{H^{-1}}. \tag{6.4}$$

Moreover, C_δ may be chosen to be a nondecreasing, continuous function of δ^{-1}.

Proof. Choose coordinates in which $x_0 = 0$, and let $B_r = \{x : |x| < r\}$. Since $X_1(0) \neq 0$,

$$\|u\|_{L^2(B_r)} \leq Cr\|X_1 u\| + C\|u\|_{L^2(B_{2r}\backslash B_r)} \quad \text{for all } r > 0,$$

by the fundamental theorem of calculus. For any r and any $\varepsilon > 0$,

$$\|u\|_{L^2(U\backslash B_r)} \leq \varepsilon \sum \|X_j u\| + C_{\varepsilon,r}\|u\|_{H^{-1}},$$

by the bracket condition and hypoellipticity. Putting these ingredients together and choosing $r = \delta$ gives

$$\|u\| \leq \|u\|_{L^2(B_\delta)} + \|u\|_{L^2(U\backslash B_\delta)} \leq C\delta\|X_1 u\| + C_\delta \varepsilon \sum \|X_j u\| + C_{\varepsilon,\delta}\|u\|_{H^{-1}}.$$

Choosing ε as a function of δ so that $\varepsilon C_\delta \leq \delta$ completes the proof. □

Lemma 6.2. *Suppose that a finite collection $\{X_j\}$ of C^∞ real vector fields satisfies (6.4) in some open set $U \subset \mathbb{R}^d$. Then there exists $w : \mathbb{R}^d \mapsto \mathbb{R}^+$ such that $w(\xi) \to \infty$ as $|\xi| \to \infty$, and such that for every $u \in H^1$ supported in U,*

$$\int_{\mathbb{R}^d} w^2(\xi)|\hat{u}(\xi)|^2 \, d\xi \leq C \sum_j \|X_j u\|^2 + C\|u\|^2. \tag{6.5}$$

Conversely, (6.5) easily implies (6.4), though this fact will not be needed here.

Proof. Let $\{P_t, \tilde{P}_t : t \in \mathbb{R}^+\}$ be the collection of operators employed in the proof of Lemma 3.1. Let $t \mapsto \delta(t)$ be a nonincreasing function to be chosen below, tending to zero as $t \to \infty$. Uniformly for all sufficiently large t,

$$\begin{aligned}\|P_t u\| &\leq \delta(t) \sum_j \|X_j P_t u\| + C_{\delta(t)}\|P_t u\|_{H^{-1}} \\ &\leq \delta(t)\Big(\sum \|P_t X_j u\| + C\|\tilde{P}_t u\| + Ct^{-1}\|u\|\Big) + C_{\delta(t)} t^{-1}\|P_t u\|.\end{aligned}$$

Here $C_\delta < \infty$ depends continuously on δ.

Choose $\delta(t)$ to be a nonincreasing continuous function of t tending to zero slowly enough as $t \to \infty$ that $t^{-1} \cdot C_{\delta(t)} \to 0$ and $\delta(t) \geq t^{-1/2}$. Then the final term in the preceding inequality may be absorbed into the left hand side for large t, yielding

$$\|P_t u\| \le \delta(t) \sum \|P_t X_j u\| + C\delta(t)\|\tilde{P}_t u\| + Ct^{-1}\|u\|,$$

whence

$$\|\delta(t)^{-1} P_t u\| \le \sum \|P_t X_j u\| + C\|\tilde{P}_t u\| + Ct^{-1}\delta(t)^{-1}\|u\|,$$

and moreover the last term is $\le Ct^{-1/2}\|u\|$. Squaring both sides and integrating yields

$$\int_1^\infty \|\delta(t)^{-1} P_t u\|^2 \tfrac{dt}{t} \le C \sum \|X_j u\|^2 + C\|u\|^2.$$

Because δ^{-1} is a continuous function of t which tends to ∞ as $t \to \infty$, the left hand side is $\ge \int_{\mathbb{R}^d} w^2(\xi)|\hat{u}|^2(\xi)\, d\xi - C\|u\|^2$ for some continuous function w tending to ∞ as $|\xi| \to \infty$, by Plancherel's theorem and the definition of $\{P_t\}$. □

Proof of Theorem 3.5. Suppose that $\{X_j\}$ satisfies the bracket hypothesis at every point of $V\backslash\{x_0\}$, and that $X_j(x_0) \neq 0$ for some j. Let $U \Subset V$ be any sufficiently small relatively compact neighborhood of x_0. Fix a nonnegative function $\Psi \in C^\infty(U)$ such that $\Psi \equiv 0$ in some small neighborhood of x_0, yet $\Psi \equiv 1$ except in a relatively compact subset of U. Let $\eta' \in C^\infty(U)$ be any function supported in the region where $\Psi \equiv 1$, such that $\eta' \ge c > 0$ except on a compact subset of U. Define the function $\psi(x, \xi)$ to be $\Psi(x)$.

For any $\delta > 0$ there exists C_δ such that for any $u \in C_0^\infty(U)$, for all i, the principal hypothesis

$$\|\operatorname{Op}[\log\langle\xi\rangle\{\psi, \sigma(X_i)\}]u\| \le \delta \sum_j \|X_j u\|^2 + C_\delta\|u\|^2$$

of Theorem 2.4 holds because $\{X_j\}$ satisfies a subelliptic estimate on a neighborhood of the set of all x for which there exists ξ such that $\{\psi, \sigma(X_i)\}(x, \xi) \neq 0$, and because of the pseudolocality of operators in the class $\operatorname{Op}(S^m_{1-\varepsilon,\varepsilon})$. Theorem 2.4 therefore applies.

The proof of the second part of the theorem proceeds in the same manner but relies instead on Theorem 2.3. The details are left to the reader, as is the proof of Theorem 3.7. □

The relevant literature is extensive. The following is a list of works cited in the text, plus a few others that are particularly relevant to C^∞ hypoellipticity. I apologize to those authors whose works are relevant but are not listed.

References

[1] D. Bell, Degenerate Stochastic Differential Equations and Hypoellipticity, Longman, Essex, 1995.

[2] D. Bell and S. Mohammed, An extension of Hörmander's theorem for infinitely degenerate differential operators, Duke Math. J. 78 (1995), 453-475.

[3] H. Boas and E. Straube, Sobolev estimates for the $\bar{\partial}$-Neumann operator on domains in $\mathbb{C}^n$ admitting a defining function that is plurisubharmonic on the boundary, Math. Z. 206 (1991), 81–88.

[4] P. Bolley, J. Camus and J. Nourrigat, La condition de Hörmander–Kohn pour les opérateurs pseudo-différentiels, Comm. Partial Differential Equations 7 (1982), 197–221.

[5] A. Bove and D. Tartakoff, Optimal non-isotropic Gevrey exponents for sums of squares of vector fields, Comm. Partial Differential Equations 22 (1997), 1263–1282.

[6] D.-C. Chang, A. Nagel and E. M. Stein, Estimates for the $\bar{\partial}$-Neumann problem in pseudoconvex domains of finite type in $\mathbb{C}^2$, Acta Math. 169 (1992), 153–228.

[7] M. Christ, Intermediate optimal Gevrey exponents occur, Comm. Partial Differential Equations 22 (1997), 359–379.

[8] —, Analytic hypoellipticity in dimension two, Math. Sci. Res. Inst. Publ., preprint 1996-009.

[9] —, Examples pertaining to Gevrey class hypoellipticity, Math. Res. Lett. 4 (1997), 725–733.

[10] —, Hypoellipticity: Geometrization and Speculation, in: Complex Analysis and Geometry (Dolbeault et al., eds.), Proceedings of the International Conference in honor of P. Lelong, Progr. Math. 188, Birkhäuser, Basel 2000, 91–109.

[11] M. Derridj and C. Zuily, Sur la régularité Gevrey des opérateurs de Hörmander, J. Math. Pures Appl. 52 (1973), 309–336.

[12] —, Régularité analytique et Gevrey pour des classes d'opérateurs élliptiques paraboliques dégénérés du second ordre, Astérisque 2, 3 (1973), 371–381.

[13] V. S. Fediĭ, On a criterion for hypoellipticity, Math. USSR-Sb. 14 (1971), 15–45.

[14] C. Fefferman, The uncertainty principle, Bull. Amer. Math. Soc. 9 (1983), 129–206.

[15] V. V. Grušin, A certain class of elliptic pseudodifferential operators that are degenerate on a submanifold, Mat. Sb. 84 (1971), 163–195 = Math. USSR-Sb. 13 (1971), 155–185.

[16] A. Grigis and J. Sjöstrand, Front d'onde analytique et sommes de carrés de champs de vecteurs, Duke Math. J. 52 (1985), 35–51.

[17] L. Hörmander, Hypoelliptic second order differential equations, Acta Math. 119 (1967), 147–171.

[18] —, The Weyl calculus of pseudodifferential operators, Comm. Pure Appl. Math. 32 (1979), 359–443.

[19] K. Kajitani and S. Wakabayashi, Propagation of singularities for several classes of pseudodifferential operators, Bull. Sci. Math. 2e Sér. 115 (1991), 397–449.

[20] J. J. Kohn, Superlogarithmic estimates on pseudoconvex domains and CR manifolds, preprint.

[21] S. Kusuoka and D. Stroock, Applications of the Malliavin calculus II, J. Fac. Sci. Univ. Tokyo 32 (1985), 1–76.

[22] P. Malliavin, C^k-hypoellipticity with degeneracy, in: Stochastic Analysis (A. Friedman and M. Pinsky, eds.), Academic Press, New York 1978, 199–214; C^k-hypoellipticity with degeneracy. II, ibid., 327–340.

[23] G. Métivier, Non-hypoellipticité analytique pour $D_x^2 + (x^2 + y^2)D_y^2$, C. R. Acad. Sci. Paris 292 (1981), 401–404.

[24] T. Matsuzawa, Gevrey hypoellipticity for Grushin operators, Publ. Res. Inst. Math. Sci. 33 (1997), 775–799.

[25] Y. Morimoto, Hypoellipticity for infinitely degenerate elliptic operators, Osaka J. Math. 24 (1987), 13–35.

[26] Y. Morimoto, A criterion for hypoellipticity of second order differential operators, Osaka J. Math. 24 (1987), 651–675.

[27] Y. Morimoto and T. Morioka, The positivity of Schrödinger operators and the hypoellipticity of second order degenerate elliptic operators, Bull. Sci. Math. 121 (1997), 507–547.

[28] O. Oleĭnik and E. Radkevič, Second-order equations with nonnegative characteristic form, Amer. Math. Soc., Providence, RI, 1973.

[29] J. Sjöstrand, Singularités analytiques microlocales, Astérisque 95 (1982).

[30] —, Analytic wavefront sets and operators with multiple characteristics, Hokkaido Math. J. 12 (1983), 392–433.

[31] M. Taylor, Pseudodifferential Operators, Princeton Univ. Press, Princeton, NJ, 1981.

[32] F. Treves, Introduction to Pseudodifferential and Fourier Integral Operators, Volume 1, Plenum Press, New York 1980.

[33] —, Analytic hypo-ellipticity of a class of pseudodifferential operators with double characteristics and applications to the $\bar{\partial}$-Neumann problem, Comm. Partial Differential Equations 3 (1978), 475–642.

[34] —, Symplectic geometry and analytic hypo-ellipticity, in: Differential equations: La Pietre 1996, Proc. Sympos. Pure Math. 65, Amer. Math. Soc., Providence, RI, 1999, 201–219.

[35] N. Dencker, Y. Morimoto, and T. Morioka, Hypoellipticity for operators of infinitely degenerate Egorov type, Tsukuba J. Math. 23 (1999), 215–224.

[36] T. Gramchev and Y. Yoshino, WKB analysis to global solvability and hypoellipticity, Publ. Res. Inst. Math. Sci. 31 (1995), 443–464.

[37] Y. Morimoto, Nonhypoellipticity for degenerate elliptic operators, Publ. Res. Inst. Math. Sci. 22 (1986), 25–30.

[38] —, Erratum to: "Nonhypoellipticity for degenerate elliptic operators", Publ. Res. Inst. Math. Sci. 30 (1994), 533–534.

[39] —, Estimates for degenerate Schrödinger operators and hypoellipticity for infinitely degenerate elliptic operators, J. Math. Kyoto Univ. 32 (1992), 333–372.

[40] —, Hypoelliptic operators in $\mathbb{R}^3$ of the form $X_1^2 + X_2^2$, J. Math. Kyoto Univ. 32 (1992), 461–484.

[41] —, Estimates for degenerate Schrödinger operators and an application for infinitely degenerate hypoelliptic operators, Proc. Japan Acad. Ser. A Math. Sci. 65 (6) (1989), 155–157.

[42] —, Propagation of wave front sets and hypoellipticity for degenerate elliptic operators, Recent developments in hyperbolic equations (Pisa, 1987), 212–224, Pitman Res. Notes Math. Ser. 183, Longman Sci. Tech., Harlow 1988.

[43] Y. Morimoto and T. Morioka, Some remarks on hypoelliptic operators which are not micro-hypoelliptic, Publ. Res. Inst. Math. Sci. 28 (1992), 579–586.

[44] T. Morioka, Une perturbation des opérateurs microhypoelliptiques, Tsukuba J. Math. 22 (1998), 537–550.

[45] —, Hypoellipticité pour un certain opérateur caractéristique double, Tsukuba J. Math. 21 (1997), 739–762.

[46] —, Hypoellipticity for some infinitely degenerate elliptic operators of second order, J. Math. Kyoto Univ. 32 (1992), 373–386.

[47] M. Reissig and K. Yagdjian, On the Stokes matrix for a family of infinitely degenerate operators of second order, Tsukuba J. Math. 21 (1997), 671–706.

[48] —, An interesting connection between hypoellipticity and branching phenomena for certain differential operators with degeneracy of infinite order, Rend. Mat. Appl. (7) 15 (1995), 481–510 (1996).

[49] N. Ueki, Hypoellipticity for the $\bar{\partial}$-Neumann problem on certain domains with infinite degeneracy, Japan. J. Math. (N.S.) 22 (1996), 79–128.

Department of Mathematics
University of California
Berkeley, Ca. 94720-3840, U.S.A.
mchrist@math.berkeley.edu

Spiraling and nonhypoellipticity

Michael Christ *

1. Introduction

Consider a smooth, pseudoconvex CR structure on some small open set of real dimension three. A fundamental sufficient condition for an associated Cauchy–Riemann operator $\bar{\partial}_b$ to be C^∞ hypoelliptic, modulo its nullspace (this notion is defined below in §4) is that the CR structure should be of finite type, but this condition is far from necessary. While some sufficient conditions are known **[2]** (see also **[16]** for the higher-dimensional case), no useful complete characterization of hypoellipticity is known, or expected, in its absence.

Some sufficient conditions take a quantitative form. For example, one such condition is that the set of weakly pseudoconvex points should be contained in a manifold M of real dimension two that is everywhere transverse to the complex tangent space, and moreover the Levi form $\Lambda(z)$ should not tend to zero too rapidly as this degenerate manifold is approached; the critical threshold is distance$(z, M)|\log \Lambda(z)| \to \infty$ as $z \to M$. A more qualitative criterion is that points of infinite type should be isolated; then the rate of degeneration does not matter. It would be desirable to find other qualitative criteria. The main result of this paper is an example answering one specific question along these lines posed by J. J. Kohn.

Theorem 1.1. *There exists a smooth, pseudoconvex, embeddable three-dimensional CR structure for which the set of weakly pseudoconvex points consists of a single smooth real curve, which is everywhere transverse to the complex tangent space, yet such that $\bar{\partial}_b$ fails to be C^∞ hypoelliptic, modulo its nullspace.*

Such a result would be of modest interest, if the production of an example were simply a matter of making the Levi form degenerate sufficiently rapidly near the curve. It is not so; Theorem 1.2 states that for a certain class of such structures, hypoellipticity always holds provided that the size of the Levi form is a monotonic function of the distance to the curve, regardless of the rate at which it degenerates. And an extremely

*The author was supported in part by NSF grant DMS-9970660 and performed this research while on appointment as a Miller Research Professor in the Miller Institute for Basic Research in Science.

weak lower bound on the rate of degeneration suffices without monotonicity; see Theorem 1.3.

An established heuristic is that the formation and propagation of singularities should be closely related to complex disks located in $\mathbb{C}^2$ near the (embedded) CR manifold. The presence of a real curve γ on which the Levi form vanishes, *transverse* to the complex tangent space, might appear to have nothing to do with the presence of such disks; hence the hope for hypoellipticity. We prefer to work intrinsically on the CR manifold itself, rather than in a neighborhood of an embedded realization. One should then look for remnants of complex disks, consisting of sufficiently large pieces of manifolds of real dimension two, on which the Levi form is sufficiently small; bearing in mind that the meaning of "sufficiently" must be clarified. Our example has an infinite sequence of nearly Levi-flat open regions converging to γ; the "leaves" of these near-foliations wind around γ. An appropriate motion "tangent" to the "leaves" is microscopically transverse to γ, but is macroscopically parallel to it, as a result of repeated windings. The construction is based on an interplay between the rates of degeneration of the Levi form, and of the winding; the dangerous situation in which hypoellipticity fails is, in our example and according to the prediction of **[1]**, when one loop around γ produces a comparatively large (though extremely small) degree of parallel motion, relative to the size of the Levi form. On the other hand, Theorem 1.4 states that if winding once around γ produces a parallel motion that is sufficiently small, relative to the size of the Levi form, then hypoellipticity does hold.

We emphasize that these results, both positive and negative, are predicted by a general but conjectural point of view formulated in **[1]**. Various works related to the construction in this paper, and likewise supporting those predictions, are listed in the bibliography. However, we have no general procedure for establishing the failure of hypoellipticity when predicted. Nor are those predictions entirely precise; they are intended only as a first-order guide to the truth. In the present paper, very special CR structures with certain symmetries are painstakingly constructed so as to support singular solutions, whose structure is not entirely obvious.

We conclude by indicating a few sufficient conditions for $\bar\partial_b$ to be C^∞ hypoelliptic, modulo its nullspace. With the exception of the second part of Theorem 1.3, these results, like the example in Theorem 1.1, concern a special class of CR structures, which we call *cylindrically symmetric*; in coordinates $(r, \theta, t) \in \mathbb{R}^+ \times [0, 2\pi] \times \mathbb{R}$, the degenerate curve γ is the axis $r = 0$, and both translation with respect to t and rotation about γ are CR symmetries; the Levi form is thus a function $\Lambda(r)$ of r alone. See §2 for the precise definitions.

These theorems demonstrate (i) that the Levi form $\Lambda(r)$ can degenerate arbitrarily rapidly as $r \to 0$, without losing hypoellipticity, (ii) that (in the cylindrically symmetric case) hypoellipticity depends on an interplay between the sizes of $\Lambda(r)$ and a second quantity $\beta(r)$, rather than merely on the size of Λ alone, and (iii) that more rapid degeneration of Λ is needed near a real curve, than near a real hypersurface, to imperil hypoellipticity. These conditions are not exhaustive; more refined sufficient conditions can be obtained by the same method.

Theorem 1.2. *Consider any smooth, pseudoconvex, cylindrically symmetric three-dimensional CR structure, as described in* §2. *Suppose that the Levi form* $\Lambda(r)$ *vanishes only at* $r = 0$, *and is a nondecreasing function of* $r > 0$. *Then* $\bar{\partial}_b$ *is* C^∞ *hypoelliptic, modulo its nullspace.*

Theorem 1.3. *Consider any smooth, pseudoconvex, cylindrically symmetric three-dimensional CR structure. Suppose that there exists* $M < \infty$ *such that for all sufficiently small* r, $\Lambda(r) \geq \exp(-r^{-M})$. *Then* $\bar{\partial}_b$ *is* C^∞ *hypoelliptic, modulo its nullspace.*

More generally, consider any smooth, pseudoconvex three-dimensional CR structure for which the set of all points of infinite type forms a smooth real curve γ, *everywhere transverse to the complex tangent space. Suppose that there exists* $M < \infty$ *such that for all points* z *sufficiently close to* γ, $\Lambda(z) \geq \exp(-d(z)^{-M})$, *where* $d(z)$ *denotes the distance from* z *to* γ. *Then* $\bar{\partial}_b$ *is* C^∞ *hypoelliptic, modulo its nullspace.*

This should be contrasted with the corresponding result for the case where γ is a smooth real hypersurface; then the above condition suffices if $M < 1$, but not if $M = 1$ **[11]**.

The next variant emphasizes the interplay between the magnitudes of Λ and a second quantity, which is related to the winding motion mentioned above.

Theorem 1.4. *Consider any smooth, pseudoconvex, cylindrically symmetric three-dimensional CR structure, as described in* §2. *Let*

$$\beta(r) = r^{-1} \int_0^r \rho\Lambda(\rho)\, d\rho\,.$$

If there exists $M < \infty$ *such that*

$$\Lambda(r) \geq e^{-M/\beta(r)}$$

for all sufficiently small $r > 0$, *then* $\bar{\partial}_b$ *is* C^∞ *hypoelliptic, modulo its nullspace.*

Remark 1.5. There are two different, though related, notions of hypoellipticity for $\bar{\partial}_b$. One, natural for a discussion of the canonical (global) solution of $\bar{\partial}_b u = f$ on a closed CR manifold without boundary, is hypoellipticity relative to its nullspace. The other notion is microhypoellipticity of the second-order operator $\Box_b = \bar{\partial}_b \bar{\partial}_b^*$, in the usual sense of wavefront sets and microlocal analysis. A close examination of our example reveals that $\Box_b$ also fails to be microhypoelliptic, on the nontrivial half of its characteristic variety (where the principal symbol of $[\bar{\partial}_b, \bar{\partial}_b^*]$ is nonnegative); failure of microhypoellipticity of the first-order operator $\bar{\partial}_b$ near the other half is quite easy to establish and requires none of the intricacy of our construction. Details are left to the reader.

2. Cylindrically symmetric CR structures

Consider a smooth CR structure in a neighborhood of the origin in $\mathbb{R}^3$, for which a Cauchy–Riemann operator takes the form $\bar\partial_b = X + \imath Y$ where, in coordinates (x, y, t),

$$X = \frac{\partial}{\partial x} + \frac{\partial a}{\partial y}\frac{\partial}{\partial t} \quad \text{and} \quad Y = \frac{\partial}{\partial y} - \frac{\partial a}{\partial x}\frac{\partial}{\partial t}$$

with $a = a(x, y)$ a real-valued C^∞ function independent of t. Thus translation with respect to t is one symmetry of the CR structure. Any such structure is embeddable; both $x + \imath y$ and $t + \imath a(x, y)$ are CR functions.

Since $[X, Y] = \Delta_{x,y}a \cdot \partial/\partial_t$, the Levi form is the function $\Delta_{x,y}a$. Assume that $\Delta_{x,y}a \geq 0$, and moreover that $\Delta_{x,y}a = 0$ if and only if $(x, y) = 0$. Then the CR structure is pseudoconvex, and the set of weakly pseudoconvex points consists of the single real curve $\gamma = \{(0, 0, t) : t \in \mathbb{R}\}$, which is everywhere transverse to the complex tangent space $T^{1,0} \oplus T^{0,1}$. We will choose a to vanish to infinite order at the origin.

Introduce "polar" coordinates $x = r\cos(\theta)$, $y = r\sin(\theta)$ with $r \geq 0$. Assume

$$a(x, y) = b(r), \tag{2.1}$$

and that $b \in C^\infty([0, 1])$, and that b vanishes to infinite order at $r = 0$. Since $\Delta_{x,y}a = b''(r) + r^{-1}b'(r)$, the pseudoconvexity hypothesis becomes $b'' + r^{-1}b' > 0$ for all $r > 0$. Rotation about the axis $r = 0$ is a second CR symmetry.

The quantity $b'(r)$ was called $\beta(r)$ in the statement of Theorem 1.4, and can be expressed in terms of the Levi form:

$$b'(r) = r^{-1}\int_0^r \rho\Lambda(\rho)\, d\rho\,.$$

Alternatively, $\Lambda(r) = \beta'(r) + r^{-1}\beta(r)$.

Consider the vector field that, in coordinates (r, θ, t), equals $\cos(\theta)Y - \sin(\theta)X$. Its integral curves wind around γ; the radius r and hence the size of the Levi form Λ, the rate of change of θ, and the velocity $\beta(r) = b'(r)$ in the t direction all remain constant. This is the spiraling motion of our title.

3. Coordinates and separation of variables

Pass to coordinates (s, θ, t) where

$$s = \log r.$$

Set

$$b(r) = \Phi(s).$$

Since $b''(r) + r^{-1}b'(r) = e^{-2s}\Phi''(s)$, the pseudoconvexity hypothesis becomes

$$\Phi''(s) > 0 \qquad \text{for all } s \in (-\infty, 0]. \tag{3.1}$$

Our aim is to construct Φ such that $\bar{\partial}_b$ fails to be C^∞ hypoelliptic, in an appropriate sense to be specified below. Among the properties of the convex function Φ will be that $\Phi(s) > 0$ and $\Phi'(s) > 0$ for all s. The infinite order vanishing of b at $r = 0$ is equivalent to

$$\frac{d^k\Phi}{ds^k} = O(e^{Ns}) \qquad \text{as } s \to -\infty, \tag{3.2}$$

for all k, $N \geq 0$. Φ will be constructed to satisfy (3.2), thus guaranteeing that $a \in C^\infty$. This also ensures that $\Delta_{x,y}a(0) = 0$, so that the CR structure is merely weakly pseudoconvex on the axis $x = y = 0$.

Choose a Hermitian structure for which $\bar{\partial}_b^* = -(X - \imath Y)$. Then $-\Box_b = -\bar{\partial}_b\bar{\partial}_b^* = (X + \imath Y) \circ (X - \imath Y)$.

The two CR symmetries permit a separation of variables in solving $\Box_b u = 0$; we consider functions $u(x, y, t) = \exp(\imath\tau t + \imath k\theta) f(s)$. For any $k \in \mathbb{N}$ and $\tau \in \mathbb{C}$,

$$-e^{-\imath\tau t - \imath k\theta}\Box_b(e^{\imath\tau t + \imath k\theta} f(s)) = \mathcal{L}_{k,\tau} f(s) \tag{3.3}$$

where

$$\mathcal{L}_{k,\tau} = e^{-2s}\big(\partial_s - k + \tau\Phi'\big)\big(\partial_s + k - \tau\Phi'\big), \tag{3.4}$$

and the coordinates (s, θ, t) are related to (x, y, t) as above. Defining likewise

$$\bar{d}_b^* f(s) = e^{-\imath\tau t - \imath k\theta}\bar{\partial}_b^*(e^{\imath\tau t + \imath k\theta} f(s)), \tag{3.5}$$

one has

$$\bar{d}_b^* = e^{-s}(\partial_s + k - \tau\Phi')\,. \tag{3.6}$$

This operator also depends on the parameters k, τ, though this dependence is not made explicit in the notation.

4. Hypoellipticity and *a priori* inequalities

$\bar{\partial}_b$ is said to be hypoelliptic modulo its nullspace, in an open set V_0, if whenever u and $\bar{\partial}_b^* u$ belong to L^2 and $\bar{\partial}_b\bar{\partial}_b^* u \in C^\infty$ in some open subset $V \subset V_0$, then likewise $\bar{\partial}_b^* u \in C^\infty(V)$. The naturality of this notion is made apparent in **[15]** and **[13]**: on a compact, embeddable CR manifold without boundary, for any $f \in L^2$ belonging to the range of $\bar{\partial}_b$, there exists a solution $g \in L^2$ orthogonal to the nullspace of $\bar{\partial}_b$ of $\bar{\partial}_b g = f$; moreover, g may be expressed as $\bar{\partial}_b^* u$ for some $u \in L^2$.

It is convenient to disprove hypoellipticity by an argument by contradiction. If $\bar{\partial}_b$ is hypoelliptic modulo its nullspace in an open set, then the following *a priori* inequalities must hold: For any open subset V, any relatively compact subset V', and any α, there must exist $M, C < \infty$ such that for any $u \in C^\infty(V)$,

$$\|\partial^\alpha_{x,y,t}\bar\partial_b^* u\|_{C^0(V')} \le C\|u\|_{C^0(V)} + C\|\bar\partial_b^* u\|_{C^0(V)} + C\|\bar\partial_b\bar\partial_b^* u\|_{C^M(V)}. \tag{4.1}$$

It is a consequence of the Baire category theorem that hypoellipticity modulo the nullspace implies such families of inequalities.[1]

Restricting attention to functions $u(x, y, t) = \exp(\imath\tau t + \imath k\theta) f(s)$, a consequence would be that for any $\epsilon > 0$ there exist $M, C < \infty$ such that for any $\tau \in \mathbb{C}$, $k \in \mathbb{N}$, and for any function $f(s)$ such that $\exp(\imath\tau t + \imath k\theta) f(s)$ is a C^∞ function of $(x, y, t) \in V$, for all $s \le -1$,

$$|\tau||\bar d_b^* f(s)| \le Ce^{\epsilon|\operatorname{Im}(\tau)|}\|f\|_{C^0} + Ce^{\epsilon|\operatorname{Im}(\tau)|}\|\bar d_b^* f\|_{C^0} + Ce^{\epsilon|\operatorname{Im}(\tau)|}\sum_{|\beta|\le M}\|(e^{-s}\partial_s, e^{-s}k, |\tau|)^\beta \mathcal{L}_{k,\tau} f\|_{C^0}. \tag{4.2}$$

This arises from (4.1) by considering a neighborhood V of the origin in which $|t| < \epsilon$. If we disprove (4.2), we will actually have proved that $\Box_b u \in C^\infty$ does not even imply[2] that $\bar\partial_b^* u \in C^1$, let alone C^∞.

We will always work with τ having $\operatorname{Re}(\tau) > 0$. It will be essential for the construction to have $\operatorname{Im}(\tau) \ne 0$. A certain tradeoff results, since the factors $\exp(\epsilon|\operatorname{Im}(\tau)|)$ on the right-hand side of (4.2) can then potentially become large. In order to have any reasonable hope of violating (4.2), we must thus ensure that $|\operatorname{Im}(\tau)| \le C\log|\tau|$.

5. Organization of the construction

Before embarking on the details, we outline the construction, oversimplifying some details in order to focus on the main line. Φ will take the form

$$\Phi = w + \sum_{j=1}^{\infty} \psi_j \tag{5.1}$$

where ψ_j, w are nonnegative, convex functions of s, all of whose derivatives tend to zero faster than any power of $\exp(s)$ as $s \to -\infty$, and the series will converge rapidly together with all the differentiated series, so that $\Phi(s)$ will define a C^∞, radial function $a(x, y)$. The term w will be strictly convex, and will tend to zero with enormous rapidity; its only purpose is to guarantee strict pseudoconvexity where $(x, y) \ne 0$, without significantly affecting either side of the *a priori* hypoellipticity inequalities (4.2) for the trial functions under consideration.

[1] Alternatively, it is straightforward to verify that an appropriate infinite linear combination u of the sequence of trial functions constructed below satisfies $\Box_b u \in C^\infty$ and $u, \bar\partial_b^* u \in L^2$, but $\bar\partial_b^* u \notin C^\infty$.

[2] In fact, the analysis shows that $\bar\partial_b^* u$ need not be Hölder continuous of any positive order. It is likely that arbitrarily weak types of regularity could be disproved, by refining the choices of parameters in §§6,7.

The functions ψ_j will be constructed by induction on j. For each $n \in \mathbb{N}$, parameters σ, k, τ and a function f will be chosen so that $\mathcal{L}^{\Phi_n}_{k,\tau} f \equiv 0$, where $\mathcal{L}^{\Phi_n}_{k,\tau}$ is defined by replacing Φ by $\Phi_n = \sum_{j=1}^{n} \psi_j$ in the definition of $\mathcal{L}_{k,\tau}$. Thus $\mathcal{L}_{k,\tau} f$ will not vanish identically, so the last of the three terms on the right-hand side of (4.2) will be nonzero. In order to make the contribution of $\psi_{n+1}, \psi_{n+2}, \ldots$ to $\mathcal{L}_{k,\tau} f$ small, we will make those functions themselves sufficiently small, at future steps of the induction.

The most delicate role in the construction falls to (essentially but not exactly) $\operatorname{Im}(\tau)$, a tuning parameter governing the behavior of f. It will be chosen to be a zero of a certain holomorphic function, and is directly analogous to the nonlinear eigenvalues on which certain analyses of nonhypoellipticity in the real analytic category rest. See for example **[3, 4, 5, 6, 9]**. In those works, the existence of appropriate nonlinear eigenvalues is proved by indirect arguments. The smooth category is much less rigid than the real analytic category; in this paper we exploit that lack of rigidity by building in the nonlinear eigenvalues by hand.

6. Building blocks

In this section we introduce and analyze the basic building blocks of the construction. Choose a large parameter $\sigma > 0$, sufficiently large to satisfy various requirements to be imposed on it below. All the other parameters k, λ, ρ, ζ to be introduced below are functions of σ.

Define a very large positive integer k by[3]

$$\log \log k = \sigma \log \sigma. \tag{6.1}$$

Here we require σ to be chosen so that k is an integer; every other requirement imposed on σ will simply be that it is sufficiently large. Define a still larger positive real number λ by

$$\frac{\lambda}{\log \lambda} = k\sigma. \tag{6.2}$$

Let $\tau = \lambda + i\zeta$, where $\zeta \in \mathbb{C}$ is to be specified; ζ is not necessarily real.

Define

$$\psi(s) = \begin{cases} 0 & \text{if } s \le -5\sigma \\ k\lambda^{-1}(s + 5\sigma) & \text{if } -5\sigma \le s \le -3\sigma \\ 2k\lambda^{-1}(s + 4\sigma) & \text{if } -3\sigma \le s \le 0. \end{cases} \tag{6.3}$$

[3]There is some arbitrariness in the choice of k. An absolute requirement is that $|\operatorname{Im}(\tau)| \le C \log |\tau|$. We began by aiming to have $|\operatorname{Im}(\tau)| \sim c \log |\tau|$, then worked backwards from the analysis of $\mathcal{N}$ to derive (6.2), then worked out a value of k that would satisfy various other requirements. Beginning with a different relation between the imaginary and real parts of τ would lead to different formulas for k, λ; so far as we know, the construction should still succeed with other such relations, so long as $|\operatorname{Im}(\tau)| \le C \log |\tau|$.

Then ψ is continuous, piecewise linear, nondecreasing, and convex.

We have remarked that any counterexample to hypoellipticity must have a non-monotonic Levi form; this is something of a red herring, and a more essential condition, phrased here very loosely, is simultaneous largeness of ψ' and smallness of ψ''. Here $\Lambda(r) = e^{-2s}\psi''$ is indeed highly non-monotonic; it is infinite at $s = -3\sigma, -5\sigma$ (though it is quite small in the sense of measures), and zero elsewhere. More essentially, for $s \in [-5\sigma, -3\sigma]$, $\psi'' \equiv 0$, while $\psi' > 0$. It will be smoothed out slightly in the next section, near -5σ and -3σ, but will retain this character.

Throughout the present section, let

$$\phi = \psi. \tag{6.4}$$

In later sections, ψ will be merely the main part of ϕ, which will have other constituents.

Define

$$h(s) = ks - \tau\phi(s), \tag{6.5}$$

$$f(s) = e^{-h(s)} \int_{-\infty}^{s} e^{2h(t)}\, dt. \tag{6.6}$$

Then

$$\bar{\partial}_b^* f = e^{h(s)} \cdot e^{-s},$$

$$\mathcal{L}_{k,\tau}^{\phi} f \equiv 0.$$

One consequence of this definition is that $f(s) = (2k)^{-1}e^{ks}$ for all $s \leq -3\sigma$, whence $\exp(\imath\tau t + \imath k\theta) f(s) = (2k)^{-1}\exp(\imath\tau t)[re^{\imath\theta}]^k$ is a smooth function of (x, y, t).

Define

$$\mathcal{N}(\zeta) = \int_{-\infty}^{0} e^{2[ks-(\lambda+\imath\zeta)\phi(s)]}\, ds \tag{6.7}$$

for all $\zeta \in \mathbb{C}$. If $\mathcal{N}(\zeta) = 0$, then f admits the alternative representation

$$f(s) = -e^{-h(s)} \int_{s}^{0} e^{2h(t)}\, dt, \tag{6.8}$$

which will be essential to estimating f for $s > -3\sigma$. If $\zeta = 0$, so that $\tau = \lambda > 0$ is real, then it is easily verified that f is an increasing function of s; when λ is large, $f(s)$ will be much larger at $s = 0$ than for s near $-\infty$, and this would thwart our efforts to contradict the *a priori* inequalities (4.2). Those inequalities do hold for s in any compact set, because $\bar{\partial}_b$ is indeed hypoelliptic modulo its nullspace, away from $r = 0$.

In order to make $\mathcal{N}(\zeta) = 0$, we wish to make $\operatorname{Re}(\zeta)$ large, to introduce some oscillation and hence cancellation into the integral. However, there is a price to be paid: the larger is $|\operatorname{Re}(\zeta)|$, the larger becomes the right-hand side of (4.2), threatening to thwart our purpose.

One computes

$$\mathcal{N}(\zeta) = e^{-10k\sigma} \cdot (I + II + III) \tag{6.9}$$

where, if $z = (2k\sigma\lambda^{-1}) \cdot \zeta = 2\zeta/\log\lambda$,

$$\begin{cases} I = (2k)^{-1} \\ II = 2\sigma z^{-1} e^{-\imath z}\sin(z) \\ III = e^{-\imath 2z}(2k + \imath 2z\sigma^{-1})^{-1}(1 - e^{-6k\sigma - \imath 6z}). \end{cases} \tag{6.10}$$

If z is bounded, and bounded away from zero, then the principal term is II, since k, σ are large. Since $II = 0$ when $z = \pi$, we thus seek to prove the existence of a unique z satisfying $\mathcal{N}(\zeta) = 0$, with z in a small neighborhood of π.

Define

$$\rho = \exp(-\sigma \log\log\sigma); \tag{6.11}$$

this choice is not natural for the analysis of this section, but will be forced on us by the considerations of §7. Let B_ρ be the ball of radius ρ in $\mathbb{C}$, centered at π. ζ is a function of z; $\mathcal{N}(\zeta)$ is thereby a holomorphic function of $z \in B_\rho$. For $z \in \partial B_\rho, II = c_0\sigma(z - \pi + O(\rho^2))$, where c_0 is a nonzero constant, while $|I| = (2k)^{-1}$ and $|III| \leq Ck^{-1}$, provided that σ is sufficiently large. Since $(k\sigma)^{-1} = \sigma^{-1}\exp(-\exp(\sigma\log\sigma))$ is much smaller than ρ, Rouché's theorem guarantees that there exists a unique $\zeta \in \mathbb{C}$ satisfying

$$\mathcal{N}(\zeta) = 0 \text{ and } |\zeta - \frac{\pi}{2}\log\lambda| \leq \rho\log\lambda. \tag{6.12}$$

In particular, $|\zeta| \leq 2\log\lambda$, provided σ is sufficiently large, and

$$|\operatorname{Im}(\zeta)| \leq \rho\log\lambda \ll \log\lambda. \tag{6.13}$$

In order to analyze the validity of (4.2), we require bounds on the various terms appearing there. For all σ sufficiently large,

$$|\bar{d}_b^* f(-4\sigma)| \geq e^{-5k\sigma} \tag{6.14}$$

$$\sup_s |\bar{d}_b^* f(s)| \leq Ce^{-5k\sigma+5\sigma} \tag{6.15}$$

$$\sup_s |f(s)| \leq C\sigma e^{-5k\sigma} \tag{6.16}$$

where $C < \infty$ is independent of σ.

In justifying these, it will be useful to know that the factor $\exp(-\imath\zeta\phi(s))$ is bounded uniformly in s, σ; in fact, its absolute value approaches 1 as $\sigma \to \infty$. Indeed, $\phi(s) \equiv 0$ for $s \leq -5\sigma$, and for $-5\sigma \leq s \leq -3\sigma$, the real part of the exponent equals $k\lambda^{-1}(s + 5\sigma)\operatorname{Im}(\zeta)$, whose absolute value is $\leq Ck\sigma\lambda^{-1} \cdot \rho\log\lambda = C\rho$, which tends rapidly to 0 as $\sigma \to \infty$. For $s \geq -3\sigma$, the exponent becomes $-\imath\zeta k\lambda^{-1} \cdot 2(s + 4\sigma)$, to which the same analysis applies.

Thus $\bar{d}_b^* f = \exp(-s + ks - \lambda\phi(s)) \cdot \exp(-\imath\zeta\phi(s))$ has absolute value bounded above and below by constant multiples of $\exp(-s + ks - \lambda\phi(s))$. We have

$$e^{-s+ks-\lambda\phi(s)} = \begin{cases} e^{(k-1)s} & s \le -5\sigma \\ e^{-5k\sigma - s} & -5\sigma \le s \le -3\sigma \\ e^{-8k\sigma - ks - s} & s \ge -3\sigma\ . \end{cases}$$

The maximum value is $\exp(-5k\sigma + 5\sigma)$, assumed at $s = -5\sigma$. For $s = -4\sigma$, this last quantity equals $\exp(-5k\sigma + 4\sigma)$. This establishes the first two of the above three bounds.

Lastly, consider f itself. By virtue of the uniform boundedness of $\exp(-\imath\zeta\phi(s))$,

$$|f(s)| \le \tilde{f}(s) = e^{-ks+\lambda\phi(s)} \int_{-\infty}^{s} e^{2(kt-\lambda\phi(t))}\,dt\ . \tag{6.17}$$

Now $kt - \lambda\phi(t)$ is a nondecreasing function of t in $(-\infty, -3\sigma]$, so

$$\tilde{f}(s) \le \int_{-\infty}^{s} e^{kt-\lambda\phi(t)}\,dt \qquad \text{for all } s \le -3\sigma.$$

Thus for $s \le -5\sigma$, $\tilde{f}(s) \equiv k^{-1}\exp(ks) \le \exp(-5k\sigma)$. For $s \in [-5k\sigma, -3k\sigma]$,

$$\begin{aligned} \int_{-\infty}^{s} e^{kt-\lambda\phi(t)}\,dt &= k^{-1}e^{-5k\sigma} + \int_{-5\sigma}^{s} e^{-5k\sigma}\,dt \\ &= k^{-1}e^{-5k\sigma} + (s+5\sigma)e^{-5k\sigma} \le (2\sigma+1)e^{-5k\sigma}\ , \end{aligned}$$

whence $\tilde{f}(s) \le (2\sigma + 1)\exp(-5k\sigma)$. Lastly, for $s \ge -3\sigma$, we invoke the alternative representation (6.8) to majorize $|f(s)|$ in the same way by

$$\begin{aligned} Ce^{-ks+2k(s+4\sigma)} \int_{s}^{0} e^{2[kt-2k(t+4\sigma)]}\,dt &\le C\int_{s}^{0} e^{-kt-8k\sigma}\,dt \\ &\le Ck^{-1}e^{-8k\sigma}e^{ks} \le Ce^{-5k\sigma}. \end{aligned}$$

This concludes the analysis of our basic building blocks.

7. Two modifications

The construction in §6 suffers from two defects. Firstly, ψ is not smooth. Secondly, we will be unable to make the final inductive scheme succeed, without taking into account in the construction of each building block ψ_j all preceding steps of the construction. In this section, we modify the construction to remedy both these defects.

Consider a mollification $\tilde{\psi}$ of ψ; define $\tilde{\psi}(s)$ to equal $\psi(s)$ unless[4] $|s+5\sigma| \le \rho$ or $|s+3\sigma| \le \rho$. In those two intervals, modify ψ so that $\tilde{\psi}$ is convex and smooth, $\tilde{\psi} \ge \psi$, $\tilde{\psi} \le \psi + 2k\lambda^{-1}|s+5\sigma|$ or $\le \psi + 2k\lambda^{-1}|s+3\sigma|$, respectively, $0 \le \tilde{\psi}'(s) \le 2k\lambda^{-1}$ for all s, and $\tilde{\psi}$ satisfies natural bounds in those intervals:

$$\left|\frac{d^n\tilde{\psi}}{ds^n}\right| \le C_n\rho^{1-n}k\lambda^{-1} \tag{7.1}$$

for every $n \ge 0$.

We claim that

$$|\frac{d^n\psi}{ds^n}(s)| \le C_{M,n}e^{-\sigma}e^{Ms} \text{ for every } M, n, \text{ for all } s \in (-\infty, 0], \tag{7.2}$$

uniformly for all sufficiently large σ. Indeed, since

$$\begin{aligned}\rho^{1-n}k\lambda^{-1} &\le \rho^{-n}k\sigma\lambda^{-1} = \exp[n\sigma\log\log\sigma - \log\log\lambda]\\ &\le \exp[n\sigma\log\log\sigma - \log\log k] = \exp[n\sigma\log\log\sigma - \sigma\log\sigma]\\ &\le C_n\exp[-\tfrac{1}{2}\sigma\log\sigma],\end{aligned}$$

(7.1) implies that $\tilde{\psi}$ satisfies (7.2) on the two intervals where it differs from ψ. Elsewhere, $\tilde{\psi} = \psi$ is linear. Its first derivative vanishes for $s \le -5\sigma$, and elsewhere equals either k/λ or $2k/\lambda$; $k/\lambda = 1/\sigma\log\lambda \le 1/\log k \le \exp(-\sigma\log\sigma)$ satisfies the required bound. The reasoning for $n = 0$ is the same.

We claim that all the bounds of §6 remain valid, if ψ is replaced by $\tilde{\psi}$, provided that ζ is such that $z = z(\zeta) \in B_\rho$. The crux of the analysis is $\exp(10k\sigma)\cdot\mathcal{N}$. Terms I, III are now strictly smaller than before, because τ has positive real part, and $\tilde{\psi} \ge \psi$. For any $z \in B_\rho$, the new principal term II differs from the old one by no more than $C\rho$, because both the new and the old integrands in the integral defining $\mathcal{N}$ are uniformly bounded multiples of $\exp(-10k\sigma)$. Thus $|II - c_0\sigma[z - \pi + O(\rho^2)]| \le C\rho$, with C independent of σ. Therefore for all sufficiently large σ, Rouché's theorem again applies on B_ρ.

We also require the three upper and lower bounds (6.14), (6.15), (6.16) on $f, \bar{d}_b^* f$. The upper bound on $\bar{d}_b^* f$ continues to hold, because $\exp(ks - \lambda\tilde{\psi}(s)) \le \exp(ks - \lambda\psi(s))$; the lower bound at $s = -4\sigma$ is unaffected because $\tilde{\psi}(-4\sigma) = \psi(-4\sigma)$. To analyze the upper bound for $f(s)$ for $s \le -3\sigma$, we pass to $\tilde{f}$ as in (6.17), and again majorize $\tilde{f}(s) \le \exp(ks - \lambda\tilde{\psi}(s))$ for all $s \le -3\sigma$. Since $\tilde{\psi} \ge \psi$, we obtain exactly the same bound as before.

Our second modification is to consider more general phases ϕ. Take

$$\phi = \psi + \varphi, \tag{7.3}$$

where ψ now denotes the function denoted above by $\tilde{\psi}$, and φ is some function which is considered to be known at the outset, before σ is chosen. Our aim is to show that if

[4] ρ here enjoys a role unrelated to its part in the preceding section; it is happenstance that the same quantity does both jobs.

σ is chosen to be sufficiently large, then all bounds obtained previously remain valid, uniformly in σ, φ, provided only that $\sigma \geq C(\varphi)$, a constant which is permitted to depend on φ. Thus ψ will depend on φ.

We assume that φ is a nonnegative, C^∞ convex function of $s \in (-\infty, 0]$, having compact support. We impose on σ the additional requirement that it be sufficiently large that the support of φ is contained in $[-\sigma, 0]$. In the analysis of $\mathcal{N}(\zeta)$, only term *III* is changed. Since $\varphi \geq 0$, the exponent $-\tau k\lambda^{-1}\varphi$ has nonpositive real part, and hence the absolute value of the integrand in this region is no larger than it was before. Therefore term *III* satisfies the same upper bound, and hence the existence proof and resulting bounds for ζ are unchanged. The real part of $\imath\zeta k\lambda^{-1}\varphi(s)$ has absolute value $\leq C\rho \log\lambda \cdot k\lambda^{-1}$, where C depends on φ; this is $\leq C\rho\sigma^{-1}$, which tends to zero rapidly as $\sigma \to \infty$. So for all σ greater than some threshold depending on φ, $\exp(\imath\zeta k\lambda^{-1}\varphi(s))$ is bounded, uniformly in s, σ, φ. Retracing the analysis of the bounds for $\bar{d}_b^* f$, f, one finds that the same estimates remain valid, provided always that σ is sufficiently large relative to φ.

8. Conclusion of the proof

Construct Φ as follows. Set $\Phi = \Psi + w$ where $\Psi = \sum_{j=1}^\infty \psi_j$ and $w = \sum_{j=1}^\infty w_j$. The functions ψ_j, w_j are constructed by induction; each ψ_j is constructed as in §6, with the two modifications of §7, and in particular is associated to a parameter σ_j. We require that $\sigma_j \geq 2^j$, just to ensure that $\sigma_j \to \infty$. We also require that w_j be nonnegative and convex, $w_{j+1}(s) \equiv 0$ for all $s < -2\sigma_j$, and $w''_{j+1}(s) > 0$ for all $s \geq -\sigma_j$.

Set $w_1 \equiv 0$. The inductive step is to construct first ψ_n, then w_{n+1}, given $\{\psi_j, w_i\}$ for all $j < n$ and $i \leq n$. Set $\varphi_n = \sum_{j=1}^{n-1}\psi_j + \sum_{i=1}^n w_i$. Choose a large parameter $\sigma_n \geq 2^n$, and in terms of it define $k_n, \lambda_n, \rho_n, \zeta_n$ and $\psi = \psi_n$ as in §§6,7. Setting $\phi = \phi_n = \psi_n + \varphi_n$, we have then a solution $f_n(s) = \exp(-h(s))\int_{-\infty}^s \exp(2h(t))\,dt$ of $\mathcal{L}_{k_n,\tau_n}^{\phi_n} f_n \equiv 0$, where $h(s) = h_n(s) = k_n s - \tau_n\phi_n(s)$, and $\tau_n = \lambda_n + \imath\zeta_n$.

Choose σ_n sufficiently large that $|d^m\psi_n(s)/ds^m| \leq 2^{-n}e^{ns}$ for all s and all $0 \leq m \leq n$, and so that[5] $|\bar{d}_b^{*\phi_n} f_n(-4\sigma_n)| \geq \exp(-5k_n\sigma_n)$, $\|\bar{d}_b^{*\phi_n} f_n\|_{C^0} \leq C\exp(-5k_n\sigma_n + 5\sigma_n)$, and $\|f_n\|_{C^0} \leq C\sigma_n\exp(-5k_n\sigma_n)$, where C is a finite constant independent of n. Here $\bar{d}_b^{*\phi_n}$ is defined by replacing Φ by ϕ_n in the definition of $\bar{d}_b^*$, just as $\mathcal{L}^{\phi_n}$ is derived from $\mathcal{L}^\Phi$. That all the preceding inequalities hold for all sufficiently large σ_n, was proved in §§6,7.

Next, construct w_{n+1} to be a nonnegative, C^∞, convex function of s, satisfying $w_{n+1}(s) \equiv 0$ for all $s < -2\sigma_n$, and $w''_{n+1}(s) > 0$ for all $s \geq -\sigma_n$. In addition, choose

[5] Throughout this section, $\bar{d}_b^*$ is always defined relative to parameters k_n, τ_n; these have been suppressed in the notation.

$e^{-ns}w_{n+1}(s)$ to have C^n norm $\le 2^{-n}$, so that the infinite series $\sum_j w_j$ converges in every C^m norm, as functions of the coordinate $r = \exp(s)$.

Other conditions must also be imposed on ψ_n, w_{n+1}, in order that the last term on the right-hand side of (4.2) should not be too large. We wish to have

$$\sum_{|\beta|\le n} \|(e^{-s}\partial_s, e^{-s}k_n, |\tau_n|)^\beta \mathcal{L}^{\Phi}_{k_n,\tau_n} f_n\|_{C^0} \le e^{-5k_n\sigma_n}. \tag{8.1}$$

Indeed, we have $\mathcal{L}^{\phi_n}_{k_n,\tau_n} f_n \equiv 0$, but require control on $\mathcal{L}^{\Phi}_{k_n,\tau_n} f_n$ since it appears on the right-hand side of (4.2). We therefore impose on each ψ_n, w_{n+1} the supplemental condition

$$\sum_{|\beta|\le r} \|(e^{-s}\partial_s, e^{-s}k_r, |\tau_r|)^\beta \big(\mathcal{L}^{\phi_n}_{k_r,\tau_r} - \mathcal{L}^{\phi_{n-1}}_{k_r,\tau_r}\big) f_r\|_{C^0} \le 2^{-n}e^{-5k_r\sigma_r} \quad \text{for each } r < n.$$

(7.2) guarantees that this holds for $\big(\mathcal{L}^{\phi_{n-1}+\psi_n}_{k_r,\tau_r} - \mathcal{L}^{\phi_{n-1}}_{k_r,\tau_r}\big) f_r$, provided that σ_n is chosen to be sufficiently large. For the contribution $\big(\mathcal{L}^{\phi_n}_{k_r,\tau_r} - \mathcal{L}^{\phi_{n-1}+\psi_n}_{k_r,\tau_r}\big) f_r$ of w_{n+1}, it suffices to directly make w_{n+1} sufficiently small in C^{n+1} norm, since $f_r \in C^\infty$ and w_{n+1} is supported in a specified compact set. Now (8.1) follows by summing over all $n' > n$.

In the same way, we can and do choose ψ_n, w_{n+1} so that for each $r < n$,

$$\|(\bar{d}_b^{*\phi_n} - \bar{d}_b^{*\phi_{n-1}}) f_r\|_{C^0} \le 2^{-n-1}e^{-5k_r\sigma_r}.$$

Here $\bar{d}_b^{*\phi_n}$ is defined with respect to k_r, τ_r, not k_n, τ_n.

Putting this all together, we have the following estimates, uniformly in n:

$$|\bar{d}_b^* f_n(-4\sigma_n)| \ge \tfrac{1}{2}e^{-5k_n\sigma_n}$$

$$\|\bar{d}_b^* f_n\|_{C^0} \le Ce^{5\sigma_n}e^{-5k_n\sigma_n}$$

$$\|f_n\|_{C^0} \le C\sigma_n e^{-5k_n\sigma_n}$$

$$\sum_{|\beta|\le n} \|(e^{-s}\partial_s, e^{-s}k_n, |\tau_n|)^\beta \mathcal{L}^{\Phi}_{k_n,\tau_n} f_n\|_{C^0} \le e^{-5k_n\sigma_n}.$$

For $f = f_n$, the left-hand side of (4.2), evaluated at $s = -4\sigma_n$, is therefore $\ge \frac{1}{2}|\tau_n|\exp(-5k_n\sigma_n)$. Since we have carefully ensured that $|\mathrm{Im}(\tau_n)| \le C\log|\tau_n|$, we may choose a fixed $\epsilon > 0$ such that $\exp(\epsilon|\mathrm{Im}(\tau_n)|) \le |\tau_n|^{1/2}$ for all n. The sum of the first two terms on the right is then $\le C|\tau_n|^{1/2}\exp(5\sigma_n)\exp(-5k_n\sigma_n)$. For any fixed M, the final term is $\le C|\tau_n|^{1/2}\exp(-5k_n\sigma_n)$. Thus (4.2) can only hold for large n if $|\tau_n| \le C|\tau_n|^{1/2}e^{5\sigma_n} + C|\tau_n|^{1/2}$. Since $|\tau_n| \sim \lambda_n \ge k_n = \exp(\exp(\sigma_n\log\sigma_n))$, this is impossible. Thus with our fixed choice of ϵ, no matter how large M is chosen to be, there exist a function f and parameters k, τ for which the *a priori* bound (4.2) does not hold. The proof is complete.

9. Proofs of positive results

We have the following general sufficient condition for hypoellipticity.

Theorem 9.1. *Consider a smooth, pseudoconvex three-dimensional CR structure, and let γ be a smooth real curve that is everywhere transverse to the complex tangent space. Suppose that the CR structure is strictly pseudoconvex at every point in the complement of γ. Suppose there exist a smooth real vector field T tangent to γ, and a real-valued function h defined in some tubular neighborhood of γ, such that $Th(z) > 0$ for all z, and such that for every $\epsilon > 0$ there exists $C < \infty$ such that for every $\tau \geq C$,*

$$\log\tau \cdot |\bar\partial_b(h)(z)| \leq \epsilon(1+\tau\Lambda(z))^{1/2}. \tag{9.1}$$

Then $\bar\partial_b$ is C^∞ hypoelliptic, modulo its nullspace.

The proof is based on the general results of **[2]**, and in outline, goes as follows. Denote by $\Gamma \subset T^*\mathbb{R}^3 \backslash (\mathbb{R}^3 \times \{0\})$ the characteristic variety associated to $\bar\partial_b$, that is, the set of all points at which the principal symbol of $\bar\partial_b$ vanishes. Γ is a line bundle over $\mathbb{R}^3$. Inside Γ is Γ_0, consisting of all fibers of Γ lying over the curve γ on which the Levi form vanishes; Γ_0 may equivalently be characterized as the subset of Γ on which the principal symbol of $[\bar\partial_b, \bar\partial_b^*]$ vanishes. Γ may be decomposed as a union of two ray bundles $\Gamma^\pm$, so that the principal symbol of $[\bar\partial_b, \bar\partial_b^*]$ is nonnegative in a conic neighborhood of Γ^+, and nonpositive in a conic neighborhood of Γ^-. Set $\Gamma_0^\pm = \Gamma_0 \cap \Gamma^\pm$.

Outside Γ, both $\bar\partial_b\bar\partial_b^*$ and $\bar\partial_b$ are microhypoelliptic, since they are elliptic. Moreover, as is well known, $\bar\partial_b\bar\partial_b^*$ is microhypoelliptic in any conic open subset of the cotangent bundle disjoint from Γ_0^-, while $\bar\partial_b$ is microhypoelliptic in any conic open subset of the cotangent bundle disjoint from Γ_0^+.

Let us fix coordinates $(p; q) = (x, y, t; \xi, \eta, \tau)$ in $T^*\mathbb{R}^3$, so that $\gamma = \{(x, y, t) : (x, y) = 0\}$, $\bar\partial_b$ belongs to the span of ∂_x, ∂_y at each point of γ, and (ξ, η, τ) is dual to (x, y, t). Then $\Gamma_0 = \{(0, 0, t; 0, 0, \tau) : \tau \neq 0\}$. All functions discussed below are assumed to be supported in some small fixed neighborhood of γ, or of the cotangent bundle of such a neighborhood, as appropriate.

Under the hypotheses that the Levi form vanishes only on a curve γ transverse to the complex tangent space, there exists a (nonelliptic) symbol $w(q) \in S^1_{1,0}$ such that $w(q) \to +\infty$ as $|q| \to \infty$ and for any C^1 function u,

$$\|w(D)u\|^2 \leq C\|\bar\partial_b u\|^2 + C\|\bar\partial_b^* u\|^2 + C\|u\|^2, \tag{9.2}$$

where each norm is an L^2 norm. Here $w(D)$ denotes the pseudodifferential operator with symbol $w(q)$. This is a simple consequence of two facts. Firstly, for all u supported in the complement of any fixed neighborhood of γ, such an inequality holds, by subellipticity. Secondly, there is a Poincaré inequality: If U_δ is a tubular neighborhood of γ of width δ, then for any u,

$$\|u\|^2 \leq C\delta^2\|\bar\partial_b u\|^2 + C\delta^2\|\bar\partial_b^* u\|^2 + C\|u\|^2_{L^2(\mathbb{R}^3\backslash U_\delta)}\,.$$

It was shown in [2] that $\bar\partial_b$ is hypoelliptic modulo its nullspace, whenever it is sufficiently strong in the sense that an inequality (9.2) holds, with a symbol satisfying $w(q)/\log|q| \to \infty$ as $|q| \to \infty$. That condition holds if $\text{distance}(p,\gamma)|\log\Lambda(p)| \to 0$ as $p \to \gamma$; but we are now interested in establishing hypoellipticity in circumstances where Λ may vanish much more rapidly.

Combining this information with Theorems 2.3 and 2.4 of [2] and their proofs[6], plus the microlocalization procedure introduced by Kohn [13], plus Gårding's inequality, we conclude that in order to prove that $\bar\partial_b$ is C^∞ hypoelliptic, modulo its nullspace, it establish the existence of certain auxiliary functions ψ. Given two distinct points $p_0, p_1 \in \gamma$, we seek an admissible C^∞ function ψ defined on $\mathbb{R}^3$, such that $\psi \equiv j$ in a neighborhood of p_j for $j = 0, 1$. Admissibility means that the pseudodifferential operator Ψ with symbol $\psi(p)\log(1+|q|^2)$ satisfies for every $\delta > 0$

$$\|[\Psi, \bar\partial_b]u\|^2 + \|[\Psi, \bar\partial_b^*]u\|^2 \le \delta\|\bar\partial_b u\|^2 + \delta\|\bar\partial_b u\|^2 + C_\delta\|u\|^2$$

for all $u \in C^1$ supported in a fixed open neighborhood of γ, and where all norms are L^2 norms. By Gårding's inequality and the microlocalized analysis in [13], this operator inequality would follow from the symbol condition

$$\begin{aligned}\log|\tau||\{\psi, \sigma(\bar\partial_b)\}(p,q)| &+ \log|\tau||\{\psi, \overline{\sigma(\bar\partial_b)}\}(p,q)| \\ &\le \delta|\sigma(\bar\partial_b)(p,q)| + \delta|\sigma([\bar\partial_b, \bar\partial_b^*])(p,q)|^{1/2} + C_\delta \quad \text{for all } \delta > 0, \quad (9.3)\end{aligned}$$

where $\sigma(T)$ denotes the principal symbol of a pseudodifferential operator T. The notation $\{a, b\}$ denotes the Poisson bracket of two symbols $a(p,q)$, $b(p,q)$. In certain proofs of hypoellipticity, a careful choice of ψ (or of a symbol $\psi(p,q)$) is essential, but in all the theorems stated above, any reasonable choice of ψ turns out to work and we will simply take ψ to be identically equal to a function of t alone in a small neighborhood of γ.

Now to complete the proof of Theorem 9.1 it suffices to take the auxiliary function ψ in the above discussion to have the form $g \circ h$, for an appropriate choice of g, where h is given in the hypotheses of Theorem 9.1.

Consider now the cylindrically symmetric case. Dropping the unhelpful term $\delta|\sigma(\bar\partial_b)|$ from the right-hand side of (9.3), our condition on $\psi(t)$ becomes simply

$$\log|\tau| \cdot b'(r) \le \delta|\tau|^{1/2}\Lambda(r)^{1/2} + C_\delta$$

for all $r > 0, \tau \ge 1$, for all small $\delta > 0$. Fixing a constant M depending on δ, this inequality holds automatically when $b'(r)\log|\tau| \le M$, and also when $|\tau|$ is less than any preassigned quantity. Thus what we need to show is that $\sup_{r,\tau} \frac{\log|\tau|}{|\tau|^{1/2}} \cdot \frac{b'(r)}{\Lambda(r)^{1/2}} \to 0$ as $A \to \infty$, where the supremum is taken over all pairs r, τ such that $|\tau| \ge A$ and $\log|\tau|b'(r) \ge M$. For fixed r, the supremum over all relevant τ occurs when

[6] Those theorems are formulated for second-order operators, such as $\bar\partial_b\bar\partial_b^*$; here we also require the analogues for first-order operators, such as $\bar\partial_b$ (microlocally near Γ_0^-). The proofs of such analogues are parallel to the second-order case, with no additional complications.

$|\tau| = \exp(M/b'(r))$; then

$$\frac{\log|\tau|}{|\tau|^{1/2}} \cdot \frac{b'(r)}{\Lambda(r)^{1/2}} = \frac{M}{\Lambda(r)^{1/2} e^{M/2b'(r)}} \cdot$$

Hence a sufficient condition for the existence of an auxiliary function ψ having the desired property is simply that there exist $M < \infty$ such that

$$\Lambda(r) \cdot \exp(M/b'(r)) \to \infty \text{ as } r \to 0 \ .$$

This completes the proof of Theorem 1.4.

To deduce the first part of Theorem 1.3 from Theorem 1.4, it suffices to note that $b'(r) = O(r^n)$ for every $n < \infty$. As for Theorem 1.2, when $\Lambda(r) = r^{-1}(rb'(r))'$ is a monotonic function, we have $b'(r) = r^{-1} \int_0^r s\Lambda(s)\,dx \leq r^{-1}\Lambda(r) \int_0^r s\,ds = \frac{1}{2} r\Lambda(r)$, so the hypothesis of Theorem 1.4 is satisfied.

The general case of Theorem 1.3 can be obtained from the same arguments. Choose coordinates (x, y, t) in which $\gamma = \{(0, 0, t)\}$, and in which the real and imaginary parts X, Y of $\bar\partial_b$ lie in the span of ∂_x, ∂_y modulo $O(r^N)$ for every N, where still $r^2 = x^2 + y^2$. This may be done by parametrizing γ as $\{\gamma(t)\}$, then introducing exponential coordinates $\exp(xX + yY)(\gamma(t))$. Then the above analysis applies, with no essential changes.

Remark 9.2. These arguments, based on the fact that the quadratic form $Q(u, u) = \|\bar\partial_b u\|^2 + \|\bar\partial_b^* u\|^2 + \|u\|^2$ dominates the right-hand side of (9.1), have not exploited the full strength of Q. To go further, one could for instance exploit the uncertainty principle by arguing that because of the Poincaré-type inequality for tubular neighborhoods of γ used above, $Q^{1/2}$ dominates any pseudodifferential operator whose symbol belongs to $S^1_{1,0}$ and is $\leq c \max_r \min\{r^{-1}, |\tau|^{1/2} \min_{\rho \geq r} \Lambda(\rho)^{1/2}\}$; thus it suffices to have $b'(r) \log|\tau|$ dominated by this quantity. Still sharper lower bounds can be obtained by pushing this reasoning to its natural conclusion, estimating a lower bound for $\partial_s + (k - \tau\phi'(s))$ in $L^2((-\infty, 0], ds)$.

Remark 9.3. After this paper was completed, we discovered an additional idea which might potentially simplify the analysis and be applicable to a wide class of cylindrically symmetric structures, obviating the need for so much care in the construction. This approach has been worked out for tubular structures, degenerating along a single hyperplane, in **[11]**. As of this writing, its application to the cylindrically symmetric case is not yet complete.

References

[1] M. Christ, Hypoellipticity: Geometrization and speculation, Progr. Math. 188 (2000), 91–109, Birkhäuser, Basel.

[2] —, Hypoellipticity in the infinitely degenerate regime, in: Complex Analysis and Geometry, Ohio State Univ. Math. Res. Inst. Publ. 9. Walter de Gruyter, Berlin–New York 2001, 59–84.

[3] —, Certain sums of squares of vector fields fail to be analytic hypoelliptic, Comm. Partial Differential Equations 16 (1991), 1695–1707.

[4] —, Analytic hypoellipticity, representations of nilpotent groups, and a nonlinear eigenvalue problem, Duke Math. J. 72 (1993), 595–639.

[5] —, A necessary condition for analytic hypoellipticity, Math. Res. Lett. 1 (1994), 241–248.

[6] —, The Szegö projection need not preserve global analyticity, Ann. of Math. 143 (1996), 301–330.

[7] —, Global C^∞ irregularity in the $\bar{\partial}$-Neumann problem for worm domains, J. Amer. Math. Soc. 9 (1996), 1171–1185.

[8] —, A progress report on analytic hypoellipticity, in: Geometric Complex Analysis (J. Noguchi et. al., eds.), World Scientific, Singapore 1996, 123–146.

[9] —, Analytic hypoellipticity in dimension two, Math. Sci. Res. Inst. Publ., preprint, February 4, 1996.

[10] —, Examples pertaining to Gevrey hypoellipticity, Math. Res. Lett. 4 (1997), 725–733.

[11] —, Hypoellipticity of the Kohn Laplacian for three-dimensional tubular CR structures, preprint, September 12, 2000.

[12] M. Christ and D. Geller, Counterexamples to analytic hypoellipticity for domains of finite type, Ann. of Math. 135 (1992), 551–566.

[13] J. J. Kohn, Estimates for $\bar{\partial}_b$ on pseudoconvex CR manifolds, in: Pseudodifferential operators and applications (F. Treves, ed.) Proc. Sympos. Pure Math. 43, Amer. Math. Soc., Providence, RI, 1985, 207–217.

[14] —, Microlocal analysis on three-dimensional CR manifolds, in: Geometry and complex variables (Bologna, 1988/1990), Lecture Notes in Pure and Appl. Math. 132, Dekker, New York 1991, 229–238.

[15] —, The range of the tangential Cauchy–Riemann operator, Duke Math. J. 53 (1986), 525–545.

[16] —, Superlogarithmic estimates on pseudoconvex domains and CR manifolds, preprint.

Department of Mathematics
University of California
Berkeley, Ca. 94720-3840, U.S.A.
mchrist@math.berkeley.edu

Positivity conditions for real-analytic functions

*John P. D'Angelo**

Contents

1. Introduction

Because the complex numbers are not ordered, the study of complex analysis requires real-valued functions in making estimates. This simple point leads one to study, depending on the context, different positivity conditions for real-valued functions of several complex variables. In this paper we discuss seven positivity conditions for real-analytic real-valued functions on complex Euclidean space and the relationships among them.

Section 2 is concrete and explicit. We define the positivity conditions and prove Theorem 2.13, which provides a precise answer to the interrelationships among these conditions. We show that most of the reverse implications fail, by giving explicit counterexamples. In Proposition 2.10 we give a one-parameter family of polynomials, and we determine the values of the parameter for which the conditions hold. Section 3

*partially supported by NSF grant DMS 99-70024

considers the special case of bihomogeneous polynomials that are positive away from the origin; in this case a decisive converse assertion holds. See Theorem 3.1. In that section we also give an improvement of a result in **[D3]**. We show that a bihomogeneous polynomial, assumed to be positive on the zero set of another bihomogeneous polynomial, agrees with a quotient of squared norms of holomorphic polynomials on that set. See Theorem 3.6.

One of our seven positivity conditions is that a real-analytic function be the quotient of squared norms of holomorphic polynomial mappings. Another is the stronger condition that it be the squared norm of a holomorphic mapping. This particular class of functions arises in many applications, and we discuss these in the rest of the paper. In Sections 4 and 5 we give some applications to proper holomorphic mappings between balls. In Theorem 4.1 we use these ideas to derive the following result. Suppose $q : \mathbb{C}^n \to \mathbb{C}$ is a holomorphic polynomial that does not vanish on the closed unit ball. Then there is an integer N and a holomorphic polynomial mapping $p : \mathbb{C}^n \to \mathbb{C}^N$ such that $\frac{p}{q}$ is a rational proper mapping between B_n and B_N, and $\frac{p}{q}$ is reduced to lowest terms. The reader should observe that this is easy when $n = 1$, where one can take $N = 1$ also. Even for $n = 2$ and q of degree two, the minimum target dimension N can be arbitrarily large. We close the paper by stating without proof an isometric imbedding theorem for holomorphic bundles, proved by Catlin and the author, that arose from generalizing a result about positivity conditions for bihomogeneous polynomials.

2. Positivity conditions

Suppose that $r : \mathbb{C}^n \times \mathbb{C}^n \to \mathbb{C}$ is a holomorphic function such that, for all z, $r(z, \overline{z})$ is real. We will be concerned with both $r(z, \overline{z})$ and with $r(z, \overline{w})$, so we use the phrase *on the diagonal* when we are thinking of the real-analytic function $z \to r(z, \overline{z})$.

We put the usual complex Euclidean norms on all finite-dimensional complex vector spaces. Thus when h is a holomorphic function taking values in $\mathbb{C}^N$, we write

$$\|h\|^2 = \sum_{k=1}^{N} |h^k|^2$$

where h^k is the k-th component function. We use the same notation for the norm when h takes values in a Hilbert space, and we use $\langle \, , \rangle$ for the inner product.

We now introduce seven positivity conditions.

1) On the diagonal r is nonnegative as a function. For all z,

$$r(z, \overline{z}) \geq 0. \tag{1}$$

2) On the diagonal r is the quotient of squared norms of holomorphic mappings. Thus there is a Hilbert space $\mathcal{H}$ and holomorphic mappings f and g from $\mathbb{C}^n$ to $\mathcal{H}$

such that

$$r(z,\overline{z}) = \frac{\|f(z)\|^2}{\|g(z)\|^2}. \tag{2}$$

3) On the diagonal r is the squared norm of a holomorphic mapping. Thus there is a Hilbert space $\mathcal{H}$ and a holomorphic mapping $f : \mathbb{C}^n \to \mathcal{H}$ such that

$$r(z,\overline{z}) = \|f(z)\|^2. \tag{3}$$

4) The underlying matrix C of Taylor coefficients of r is of the form A^*A. Thus there is a Hilbert space $\mathcal{H}$ and elements $f_\alpha \in \mathcal{H}$ such that

$$r(z,\overline{z}) = \sum c_{\alpha\beta} z^\alpha \,\overline{z}^\beta$$

and the entries $c_{\alpha\beta}$ of the (necessarily) Hermitian matrix C satisfy

$$c_{\alpha\beta} = \langle f_\alpha, f_\beta \rangle. \tag{4}$$

5) There is a positive integer N such that, on the diagonal, r^N is a squared norm of a holomorphic mapping. Thus there is a Hilbert space $\mathcal{H}$ and a holomorphic mapping $f : \mathbb{C}^n \to \mathcal{H}$ such that

$$r(z,\overline{z})^N = \|f(z)\|^2. \tag{5}$$

6) r satisfies the global Cauchy–Schwarz inequality. For all z and w,

$$r(z,\overline{z})r(w,\overline{w}) \geq |r(z,\overline{w})|^2. \tag{6}$$

7) On the diagonal, r is plurisubharmonic. For all $z, a \in \mathbb{C}^n$, we have

$$\sum_{i,j=1}^{n} r_{z_i\overline{z}_j}(z,\overline{z})a_i\overline{a}_j = \partial\overline{\partial}r(z,\overline{z})(a,\overline{a}) \geq 0. \tag{7}$$

Remark 2.1. Suppose r is a polynomial. Then the only Hilbert spaces that arise are complex Euclidean spaces $\mathbb{C}^N$, for some N. Also condition 4) is then equivalent to the underlying matrix of coefficients being positive semi-definite.

Remark 2.2. Analogous considerations apply also when Ω is an open domain in complex Euclidean space, Ω^* denotes its conjugate domain, and $r : \Omega \times \Omega^* \to \mathbb{C}$ is a holomorphic function. In this paper however we generally restrict to the case when Ω is all of complex Euclidean space, because the most interesting examples will arise when r is a polynomial.

We begin by discussing the easy implications. It is immediate that 3) implies 5), that 3) implies 2), and that 2) implies 1). It is easy to see that 3) implies 7). We begin by showing that 5) implies 6). We will see below that 6) implies 7), so we will conclude that 5) implies 7). This also gives an indirect proof that 3) implies 7).

Lemma 2.3. *Statement* 5) *implies statement* 6).

Proof. Suppose that $r^N = \|f\|^2$. By the usual Cauchy–Schwarz inequality,

$$(r(z,\overline{z})r(w,\overline{w}))^N = \|f(z)\|^2\|f(w)\|^2 \geq |\langle f(z), f(w)\rangle|^2 = |r(z,\overline{w})|^{2N}. \quad (8)$$

Since $N > 0$, we may take N-th roots of both sides of (8) and preserve the direction of the inequality. □

Lemma 2.4. *The implication* 6) *implies* 1) *is false, but if* $r(z,\overline{z}) > 0$ *for a single point* z, *then* 6) *implies* 1).

Proof. First we show that when r is minus a squared norm, and not identically zero, 6) holds and 1) obviously fails. Put $r(z,\overline{z}) = -\|f(z)\|^2$ for a holomorphic mapping f. Obviously 1) fails unless f is identically 0. On the other hand, 6) holds, again by the usual Cauchy–Schwarz inequality,

$$r(z,\overline{z})r(w,\overline{w}) = \|f(z)\|^2\|f(w)\|^2 \geq |\langle f(z), f(w)\rangle|^2 = |r(z,\overline{w})|^2.$$

Suppose however that $r(z,\overline{z})$ is positive. Then 6) is equivalent to

$$r(w,\overline{w}) \geq \frac{|r(z,\overline{w})|^2}{r(z,\overline{z})} \quad (9)$$

and hence $r(w,\overline{w})$ is nonnegative for all w, so 1) holds. □

Remark 2.5. Proposition 2.10 below shows how to construct examples where 6) and 1) hold, but 3) does not.

Proposition 2.6. *Suppose for some* z *that* $r(z,\overline{z}) > 0$. *Then* 6) *implies* 7).

Proof. Suppose that 6) holds, and put $w = z + ta$ for $t \in \mathbb{C}$ and $a \in \mathbb{C}^n$. Then

$$0 \leq r(z,\overline{z})r(z+ta, \overline{z+ta}) - |r(z,\overline{z+ta})|^2 = h(t,\overline{t}). \quad (10)$$

Obviously h vanishes at $t = 0$, so this is a minimum point for the function h. Hence its Laplacian $h_{t\overline{t}}(0)$ is nonnegative. Computing the Laplacian by the chain rule and evaluating at $t = 0$ gives

$$0 \leq h_{t\overline{t}}(0) = r(z,\overline{z})\sum r_{z_j\overline{z}_k}(z,\overline{z})a_j\overline{a}_k - \left|\sum r_{z_j}(z,\overline{z})a_j\right|^2. \quad (11)$$

Restating (11) in invariant notation gives

$$r(z,\overline{z})\partial\overline{\partial}r(z,\overline{z})(a,a) \geq |\partial r(z,\overline{z})(a)|^2 \geq 0.$$

In case r is positive at z, we can divide and see that r is also plurisubharmonic there.

If r is nonnegative, and $r(z,\overline{z}) = 0$, then the function $t \to r(z+ta, \overline{z+ta})$ has a minimum at 0. Therefore its Laplacian is nonnegative at $t = 0$, and we again obtain that the complex Hessian of r at z is nonnegative. Thus, whether or not r vanishes at z, its Hessian is nonnegative there, so 7) holds. □

Corollary 2.7. *Statement* 5) *implies statement* 7). *Furthermore, suppose that h is a holomorphic mapping such that, for a real-analytic function r we have*

$$r(z,\bar{z})^N = \|h(z)\|^2.$$

Then the following inequality (as Hermitian forms) holds at each point; here h_i denotes $\frac{\partial h}{\partial z_i}$ and of course h is generally vector-valued:

$$\|h\|^2 \langle h_i, h_j \rangle \geq \left(1 - \frac{1}{N}\right) \langle h_i, h \rangle \langle h, h_j \rangle. \tag{12}$$

Proof. By Lemma 2.3, 5) implies 6). If r vanishes identically, 7) holds. Otherwise Proposition 2.6 shows that 6) implies 7). Thus 5) implies 7). Suppose now that h and r are as given. Then r is plurisubharmonic because 6) holds. First we compute the Hessian $\partial\bar{\partial} r$ using $r = (\|h\|^2)^{\frac{1}{N}}$ and the chain rule. We have

$$r_i = \frac{1}{N} (\|h\|^2)^{\frac{1}{N}-1} \langle h_i, h \rangle \tag{13}$$

and

$$r_{ij} = \frac{1}{N} (\|h\|^2)^{\frac{1}{N}-1} \langle h_i, h_j \rangle + \left(\frac{1}{N}\right)\left(\frac{1}{N} - 1\right) (\|h\|^2)^{\frac{1}{N}-2} \langle h_i, h \rangle \langle h, h_j \rangle. \tag{14}$$

Since (14) is nonnegative definite and $\|h\|^2$ takes on nonnegative values, inequality (12) follows from (14). □

We have observed that 6) implies that r is always of one sign. Choosing r to be nonnegative we can then conclude that r is actually plurisubharmonic. The converse is false; we give a simple example here, and a more subtle one in Proposition 2.10.

Example 2.8. Let $n = 1$, and put $r(z,\bar{z}) = |z|^2 + c(z^2 + \bar{z}^2)$. Then r is subharmonic for all choices of the constant c, but r changes sign when $c > 1$. Thus 6) cannot hold, so 7) does not imply 6).

We note that 6) fails for any positive c. Evaluating at $z = 1$ and $w = i$ shows that

$$|r(1,i)|^2 = 1 > 1 - 4c^2 = r(1,1)r(i,-i).$$

Thus, even when r is a homogeneous polynomial, r is nonnegative, and 7) holds, we cannot conclude that 6) holds.

In order to develop a feeling for the Cauchy–Schwarz inequality 6), let r be a real-valued polynomial r on $\mathbb{C}^n$. Then (see **[D1]**) there are holomorphic polynomial mappings f and g, taking values in a finite-dimensional space, such that

$$r(z,\bar{z}) = \|f(z)\|^2 - \|g(z)\|^2 \tag{15}$$

We may assume that there are no linear dependence relations among the components of f and g, but even then the representation is not unique. We interpret some of our positivity conditions in this setting.

Note that 1) holds if and only if $\|f(z)\|^2 \geq \|g(z)\|^2$ for all z. Condition 3) holds if and only if we may choose $g = 0$. Condition 7) holds if and only if the Hessian of $\|f\|^2$ is greater (as a matrix) than the Hessian of $\|g\|^2$. All of these assertions are obvious.

We have the following formula characterizing when 6) holds; we state it more generally.

Proposition 2.9. *Suppose that $\mathcal{H}$ is a Hilbert space, and that f and g are holomorphic mappings to $\mathcal{H}$. Put*

$$r(z, \overline{w}) = \langle f(z), f(w)\rangle - \langle g(z), g(w)\rangle. \tag{16}$$

Then 6) holds if and only if, for every pair of points z and w, we have

$$\|f(z) \otimes g(w) - f(w) \otimes g(z)\|^2 \leq \|f(z)\|^2\|f(w)\|^2 - |\langle f(z), f(w)\rangle|^2 + \|g(z)\|^2\|g(w)\|^2 - |\langle g(z), g(w)\rangle|^2. \tag{17}$$

Proof. Begin by using (16) to express 6) in terms of f and g. The resulting inequality is then seen to be equivalent to (17). To see this, expand the squared norm on the left side of (17), and use the identity $\langle u_1 \otimes v_1, u_2 \otimes v_2\rangle = \langle u_1, u_2\rangle\langle v_1, v_2\rangle$. □

Observe that each term on the right side of (17) is nonnegative by the usual Cauchy–Schwarz inequality. Our version of the Cauchy–Schwarz inequality demands that their sum bounds an obviously nonnegative expression that reveals the symmetry of the situation. The left side of (17) has the interpretation as $\|(f \wedge g)(z, w)\|^2$, but the author does not know how to use this information.

Our next proposition is important for two reasons. First it gives a concrete example showing that most of the reverse implications fail. In particular conditions 1), 3), 6), and 7) are distinct. Second, because r is bihomogeneous, the example will be useful when we think of r as giving a metric on the universal bundle over complex projective space.

Proposition 2.10. *For each real number a, define a bihomogeneous polynomial function r_a on the diagonal in $\mathbb{C}^2$ by*

$$r_a(z, \overline{z}) = \left(|z_1|^2 + |z_2|^2\right)^4 - a|z_1 z_2|^4.$$

The following then hold; (the numbering here parallels the numbering of the seven conditions).

1. *r_a is nonnegative as a function if and only if $a \leq 16$.*
2. *r_a is the quotient of squared norms of holomorphic mappings if and only if $a < 16$.*

3. *r_a is a squared norm of a holomorphic polynomial mapping if and only if $a \le 6$.*

4. *The underlying matrix of coefficients for r_a is diagonal; its eigenvalues are 1, 4, $6 - a$, 4, and 1. Thus it is positive definite if and only if $a < 6$, and positive semi-definite if and only if $a \le 6$.*

5. *r_a^2 is a squared norm if and only if $a \le 7$.*

6. *r_a satisfies the global Cauchy–Schwarz inequality for $a \le 7.8$, but does not when $a > 8$.*

7. *r_a is plurisubharmonic if and only if $a \le 12$.*

Proof. Since r_a (on the diagonal) depends upon only the absolute values of the variables, it is convenient to write $|z_1|^2 = x$ and $|z_2|^2 = y$. We then have

$$r_a(z, \bar{z}) = (x+y)^4 - ax^2y^2 = \big((x+y)^2 + \sqrt{a}xy)((x+y)^2 - \sqrt{a}xy\big). \tag{18}$$

The first factor on the far right in (18) is nonnegative, so we see that r_a is nonnegative as a function when

$$0 \le (x+y)^2 - \sqrt{a}xy = \big(x-y)^2 + (4 - \sqrt{a}\big)xy. \tag{19}$$

This implies the first statement, because when $a > 16$ we may choose $x = y$ and contradict the inequality.

Thus r_a is nonnegative as a function when $a \le 16$; it is positive away from the origin if and only if $a < 16$. Theorem 3.1 in this paper guarantees that r_a is a quotient of squared norms when $a < 16$, and Lemma 3.3 implies that it is not a quotient of squared norms when $a = 16$. We omit the details of the easy calculation using Lemma 3.3.

In terms of x and y note that $r_a = x^4 + 4x^3y + (6-a)x^2y^2 + 4xy^3 + y^4$. This proves the third and fourth statements; in particular when $a \le 6$, r_a is a squared norm. Since being a squared norm implies each of the other properties, we thus may assume that $a \ge 6$ in the rest of the proof.

It remains to prove 5), 6), and 7). We do 7) first, by using the special form of r_a to find its Hessian and to decide when it is plurisubharmonic. Writing λ_{jk} for the partial derivative $\frac{\partial}{\partial z_j}\frac{\partial}{\partial \bar{z}_k}(r_a)$ we obtain

$$\lambda_{11} = 4\big(|z_1|^2 + |z_2|^2\big)^3 + 12\big(|z_1|^2 + |z_2|^2\big)^2|z_1|^2 - 4a|z_1|^2|z_2|^4 \tag{20}$$

$$\lambda_{12} = \bar{z}_1 z_2\big(12\big(|z_1|^2 + |z_2|^2\big)^2 - 4a|z_1|^2|z_2|^2\big) \tag{21}$$

$$\lambda_{22} = 4\big(|z_1|^2 + |z_2|^2)^3\big) + 12\big(|z_1|^2 + |z_2|^2\big)^2|z_2|^2 - 4a|z_1|^4|z_2|^2. \tag{22}$$

By the bihomogeneity it suffices to determine the positive definiteness of the Hessian matrix on the unit sphere. We then obtain

$$\lambda_{11} = 4 + 12|z_1|^2 - 4a|z_1|^2|z_2|^4 \tag{23}$$

$$\lambda_{12} = \bar{z}_1 z_2 \big(12 - 4a|z_1|^2|z_2|^2\big) \tag{24}$$

$$\lambda_{22} = 4 + 12|z_2|^2 - 4a|z_1|^4|z_2|^2. \tag{25}$$

Computing the determinant $\det\lambda = \lambda_{11}\lambda_{22} - |\lambda_{12}|^2$ and eliminating $|z_2|^2$ we find that

$$\det\lambda = 64 - 64a|z_1|^2 + 256a|z_1|^4 - 384a|z_1|^6 + 192a|z_1|^8. \tag{26}$$

We know that r_a is plurisubharmonic if and only if both λ_{11} and the determinant $\det\lambda$ are nonnegative. When $a = 12$ one can verify that the determinant equals

$$64\big(1 - 6|z_1|^2 + 6|z_1|^4\big)^2 \tag{27}$$

and that $\lambda_{11} > 0$ for $0 \le |z_1|^2 \le 1$. The expression (27) vanishes at two values in the unit interval. The expression for λ_{11} is a cubic in $|z_1|^2$ with only one real zero, and it lies outside the unit interval. This shows that r_a is plurisubharmonic when $a = 12$; it follows easily that the same is true when $a \le 12$. When $a > 12$, expression (26) is negative for points close to the roots of (27). Thus $a = 12$ is the cut off point for plurisubharmonicity, and the seventh statement holds.

It remains to discuss 5) and 6). Computing r_a^2 gives

$$r_a^2 = x^8 + 8x^7y + (28-2a)x^6y^2 + (56-8a)x^5y^3 + (70-12a+a^2)x^4y^4 + \cdots . \tag{28}$$

Expression (28) is symmetric in x and y, so to make r_a^2 a squared norm we need only make the coefficients of x^6y^2, of x^5y^3, and of x^4y^4 nonnegative. From this we find that r_a^2 is a squared norm when $a \le 7$. Using Mathematica with $a = 7.8$, one computes r_a^{32} and sees that all the coefficients are positive. Thus it is a squared norm.

By Theorem 2.13, we obtain the Cauchy–Schwarz inequality 6) for $a \le 7.8$. We show next that it fails when $a > 8$ We evaluate both sides of the inequality at the points $z = \big(\frac{1}{\sqrt{2}}, \frac{1}{\sqrt{2}}\big)$ and $w = \big(\frac{1}{\sqrt{2}}, \frac{-1}{\sqrt{2}}\big)$. Then $r_a(z, \overline{w}) = \frac{-a}{16}$ and $r_a(z, \bar{z}) = r_a(w, \overline{w}) = 1 - \frac{a}{16}$. Plugging this into the inequality 6) gives the condition

$$\left|\frac{-a}{16}\right|^2 \le \left(1 - \frac{a}{16}\right)^2$$

which implies that $1 - \frac{a}{8} \ge 0$ and shows that the inequality implies that $a \le 8$. We have now proved all the statements made. □

Remark 2.11. One can generalize this example to n dimensions. The corresponding function is

$$r_a(z, \bar{z}) = \|z\|^{4n} - a\Big|\prod z_j\Big|^4.$$

Statement 1) holds for $a \le n^{2n}$. Statement 3) holds for $a \le \frac{(2n)!}{2^n}$. The Cauchy–Schwarz inequality fails for $a > \frac{n^{2n}}{2}$.

Remark 2.12. Proposition 2.10 does not provide an example where the positivity conditions numbered 5) and 6) are distinct. For r_a it follows from Theorem 6.2 (the result about isometric imbedding) that these conditions are the same; the author suspects that these conditions might be the same under general circumstances. It seems that the precise cut off point for the Cauchy–Schwarz inequality is at 8, we haven't proved that.

Next we consider the relationship between 3) and 4). These assertions are equivalent, but a warning is required in the implication that 4) implies 3).

Assume only that f_α are vectors in a Hilbert space $\mathcal{H}$ and put $c_{\alpha\beta} = \langle f_\alpha, f_\beta \rangle$. The sum $\sum c_{\alpha\beta} z^\alpha \overline{z}^\beta$ will not in general be convergent everywhere, even if the vectors f_α form an orthonormal set. Suppose for example that $n = 1$ and that $\{\phi_\alpha\}$ (for $\alpha = 0, 1, \dots$) forms an orthonormal set in $\mathcal{H}$. We let $c_{\alpha\beta} = \langle \phi_\alpha, \phi_\beta \rangle$. We then have

$$r(z, \overline{z}) = \sum c_{\alpha\beta} z^\alpha \overline{z}^\beta = \sum |z|^{2\alpha}.$$

On the open unit disk $r(z, \overline{z}) = \frac{1}{1-|z|^2}$, but obviously r is not real-analytic otherwise.

To see that 3) implies 4), we suppose that $r(z, \overline{z}) = \|f(z)\|^2$, and that the Taylor series of the Hilbert space valued function f is given by $f(z) = \sum f_\alpha z^\alpha$. Then

$$r(z, \overline{z}) = \left\| \sum f_\alpha z^\alpha \right\|^2 = \sum \langle f_\alpha, f_\beta \rangle z^\alpha \overline{z}^\beta \tag{29}$$

and hence (4) holds.

Conversely if 4) holds, we define f by $f(z) = \sum f_\alpha z^\alpha$ and see from (29) that 3) holds. The sum defining f always converges because r is assumed entire.

Suppose on the diagonal that r is a squared norm of a finite-dimensionally valued holomorphic mapping. It follows from (19) that

$$\sum c_{\alpha\beta} \zeta_\alpha \overline{\zeta}_\beta = \left\| \sum f_\alpha \zeta_\alpha \right\|^2 \geq 0 \tag{30}$$

and hence the matrix of coefficients is positive semi-definite. In case r is a polynomial, or more generally when the rank of C is finite, it is easy to see that 3) and 4) are equivalent to the assertion that C is positive semi-definite.

We summarize all these conclusions in the following result.

Theorem 2.13. *Suppose that $r : \mathbb{C}^n \times \mathbb{C}^n \to \mathbb{C}$ is a holomorphic function such that, for all z, $r(z, \overline{z})$ is real. Then the following implications hold and no others do (with the possible exception of 6) $\Rightarrow$ 5) which we do not consider here).*

(a) 3) $\Leftrightarrow$ 4).

(b) 3) $\Rightarrow$ 2) $\Rightarrow$ 1).

(c) 3) $\Rightarrow$ 5) $\Rightarrow$ 1).

(d) 5) $\Rightarrow$ 6).

Suppose in addition that $r(z, \overline{z})$ is positive at a single point z. Then

(e) 6) $\Rightarrow$ 1).

(f) 6) $\Rightarrow$ 7).

Proof. We have already proved all the implications. Example 3.4 from the Section 3 shows that 1) does not imply 2). Proposition 2.10 shows that 1) does not imply 3), 6), or 7). It shows that 7) does not imply 6), 5), or 3). It shows that 6) does not imply 3). The only implication whose truth is not established by way of these examples is the possibility that 6) implies 5), which we do not discuss here. □

3. The polynomial and bihomogeneous polynomial cases

None of the assertions in Theorem 2.13 begin with 1) as the hypothesis. It is generally false, even for bihomogeneous polynomials, that 1) implies any of the other assertions. If, however, r is a bihomogeneous polynomial and it is *strictly positive* away from the origin, then 2) holds. This result was proved in both **[CD1]** and **[Q]**. After recalling some notation we state this result here.

Suppose that

$$p(z, \overline{z}) = \sum_{|\alpha|+|\beta|\leq 2m} c_{\alpha\beta} z^{\alpha} \overline{z}^{\beta} \tag{31}$$

is a *polynomial* function on complex Euclidean space $\mathbb{C}^n$. Observe that $p(z, \overline{z})$ will be real for all z if and only if the *underlying matrix of coefficients* $C = (c_{\alpha\beta})$ is Hermitian symmetric. It is evident that, if C is non-negative definite, then the polynomial will take on non-negative values, and if C positive definite, then the polynomial will be strictly positive away from the origin. The polynomial p can be considered as the restriction of the Hermitian form in N variables

$$\sum_{\alpha,\beta=1}^{N} c_{\alpha\beta} \zeta_{\alpha} \overline{\zeta}_{\beta} \tag{32}$$

to a Veronese variety given by parametric equations $\zeta_{\alpha}(z) = z^{\alpha}$. Simple examples such as $|z_1|^4 - |z_1 z_2|^2 + |z_2|^4$ reveal that positivity conditions for the function are weaker than positivity conditions for the Hermitian form. In the language of Theorem 2.13, the implication 1) implies 4) is false. Here the Hermitian form has eigenvalues $1, -1, 1$, yet the function is positive away from the origin.

We recall some terminology and notation. Let B_n denote the unit ball in $\mathbb{C}^n$. We denote by $A^2(B_n)$ the Hilbert space of square-integrable holomorphic functions on the ball. A real-valued polynomial p on $\mathbb{C}^n$ is called *bihomogeneous of degree* $2m$ if

$$p(z, \overline{z}) = \sum_{|\alpha|=|\beta|=m} c_{\alpha\beta} z^{\alpha} \overline{z}^{\beta}. \tag{33}$$

We write V_m for the complex vector space of homogeneous holomorphic polynomials of degree m, and we can thus identify p with an Hermitian form on V_m via its underlying matrix of coefficients. In this section on polynomials we reserve the term *squared norm* for a *finite* sum of squared absolute values of holomorphic functions. Note that a power of the squared Euclidean norm is itself a squared norm; $\|z\|^{2d} = \|H_d\|^2$, where H_d is the d-fold tensor product of the identity map with itself. Observe also that the components of H_d form a basis for V_d.

There is a decisive theorem (see **[CD1, Q]**) in the bihomogeneous case: such a polynomial is positive away from the origin precisely when it is a quotient of squared norms of holomorphic polynomial mappings. This is statement 4) below.

Theorem 3.1 (Catlin–D'Angelo, Quillen). *The bihomogeneous case. Let* $f(z,\overline{z}) = \sum\sum c_{\alpha\beta} z^{\alpha}\,\overline{z}^{\beta}$ *be a real-valued polynomial that is homogeneous of degree m in z and also of degree m in* $\overline{z}$. *The following are equivalent.*

(1) *f achieves a positive minimum value on the sphere.*

(2) *There is an integer d such that the underlying Hermitian matrix for* $\|z\|^{2d} f(z,\overline{z})$ *is positive definite. Thus*

$$\|z\|^{2d} f(z,\overline{z}) = \sum E_{\mu\nu} z^{\mu} \overline{z}^{\nu} \tag{34}$$

where $(E_{\mu\nu})$ *is positive definite.*

(3) *There is an integer d such that the operator* R_{m+d} *defined by the kernel* $k_d(z,\zeta) = \langle z,\zeta\rangle^d f(z,\overline{\zeta})$ *is a positive operator from* $V_{m+d} \subset A^2(B_n)$ *to itself.*

(4) *There is an integer d and a holomorphic homogeneous vector-valued polynomial g of degree* $m+d$ *such that* $V(g) = \{0\}$ *and such that* $\|z\|^{2d} f(z,\overline{z}) = \|g(z)\|^2$. *Thus f is a quotient of squared norms.*

(5) *Write* $f(z,\overline{z}) = \|P(z)\|^2 - \|N(z)\|^2$ *for holomorphic homogeneous vector-valued polynomials P and N of degree m. Then there is an integer d and a linear transformation L such that the following are true:*

 (5.1) $I - L^*L$ *is positive semi-definite.*

 (5.2) $H_d \otimes N = L(H_d \otimes P)$.

 (5.3) $V\big(\sqrt{I - L^*L}(H_d \otimes P)\big) = \{0\}$.

We interpret Theorem 3.1 as giving a result in the language of Theorem 2.13. "For bihomogeneous polynomials, a stronger form of 1) implies 2)". Even for bihomogeneous polynomials, 1) does not imply 2). See Example 3.4 below.

The main assertion that 1 and 2 are equivalent was proved in 1967 by Quillen. Unaware of that result, Catlin and the author, motivated by trying to prove Theorem 4.3 below, found a different proof. Both proofs use hard analysis; Quillen uses Gaussian integrals on all of $\mathbb{C}^n$, whereas Catlin–D'Angelo use the Bergman kernel function on the unit ball B_n. Quillen uses a priori estimates, whereas Catlin–D'Angelo use facts about compact operators on $L^2(B_n)$. In **[CD2]** they generalized to operators on weakly pseudoconvex domains some of these facts about compact operators. In the statement of Proposition 1 of **[CD2]**, the word tangential should appear before pseudodifferential operator, but was incorrectly omitted.

In **[CD2]** they reinterpreted Theorem 2.13 in terms of holomorphic vector bundles and generalized the result considerably in **[CD3]**. We state this in Theorem 6.2 below.

A bihomogeneous polynomial that is positive away from $\{0\}$ must be the quotient of squared norms of holomorphic polynomial mappings. The generalization to non-negative bihomogeneous polynomials fails. We present two simple counterexamples from **[D3]** here.

Definition 3.2. A germ at z_0 of a real-analytic function $h : \mathbb{C} \to \mathbb{R}$ satisfies property (W) if either h is identically 0, or h vanishes to finite even order $2m$ at z_0, and its initial form $j_{2m}h$ is independent of the argument of t. Thus $j_{2m}h(t) = c|t|^{2m}$.

For example, $(2\mathrm{Re}(t))^2$ does not satisfy property (W); it equals its initial form, which is $t^2 + 2|t|^2 + \bar{t}^2$. The following simple lemma from **[D3]** reveals that property (W) is a necessary condition for being a quotient of squared norms.

Lemma 3.3. *Suppose that f, g are holomorphic mappings with domain $\mathbb{C}^n$ and taking values in a Hilbert space $\mathcal{H}$ with squared norm $\|\cdot\|^2$. Let $p(z, \bar{z}) = \frac{\|f(z)\|^2}{\|g(z)\|^2}$. Consider the germ of a holomorphic mapping $t \to z(t)$ from $\mathbb{C}$ to $\mathbb{C}^n$ such that the order of vanishing of the numerator is at least the order of vanishing of the denominator. Then the pullback function $t \to p(z(t), \overline{z(t)})$ satisfies property (W).*

Example 3.4. The following bihomogeneous polynomials are obviously non-negative but cannot be written as quotients of squared norms:

$$p(z, \bar{z}) = (|z_1|^2 - |z_2|^2)^2 \tag{35}$$

$$h(z, \bar{z}) = (|z_1 z_2|^2 - |z_3|^4)^2 + |z_1|^8. \tag{36}$$

The polynomial p from (35) is not a quotient of squared norms; its zero set is three real dimensional, and hence not a complex variety. To see that the necessary condition of Lemma 3.3 fails, we can choose $z(t) = (1 + t, 1)$. Then

$$p(z(t), \overline{z(t)}) = (t + \bar{t} + |t|^2)^2$$

and the initial form violates property (W).

The zero set of the non-negative polynomial h is the complex variety defined by $z_1 = z_3 = 0$. Yet h is not a quotient of squared norms; it doesn't satisfy the necessary condition of Lemma 2.3 either. Consider for example the mapping given by $z(t) = (t^2, 1+t, t)$. Then the pullback $h(z(t), \overline{z(t)})$ vanishes to order ten at the origin, but the terms of order ten include $t^4\bar{t}^6$.

It is tempting to conjecture that these two necessary conditions (property (W) and a complex zero set) are sufficient for writing polynomials as quotients of squared norms.

To give the reader some feeling for Theorem 3.1, we next describe a very simple version of it that can be proved by high school mathematics.

Proposition 3.5. *Let p be a polynomial in one real variable. Then $p(t) > 0$ for all $t \geq 0$ if and only if there is an integer d such that the polynomial given by $(1+t)^d p(t)$ has only positive coefficients. The minimum such d can be arbitrarily large for polynomials of fixed degree.*

Proof. (Sketch) Factor p into linear and quadratic factors. It suffices to verify the conclusion for each factor. The result is trivial for linear factors. It therefore suffices to check it for the quadratic polynomial $p(t) = t^2 + bt + c$. One can do this directly. □

Proposition 3.5, which goes back at least as far as Polya **[HLP]**, is a special case of Theorem 3.1. Replace the variable t in p by $|\frac{z}{w}|^2$ and clear denominators. The result will be a bihomogeneous polynomial f in the two complex variables z and w, positive away from the origin. It is not difficult to see that the conclusion of Theorem 3.1 amounts to the conclusion of Proposition 3.5. See also **[H]** for many related things in the real case.

Combining this with Theorem 6.2 (the reinterpretation of Theorem 3.1 in terms of bundles) shows that Proposition 3.5 can be reinterpreted as a result about metrics on the m-th power of the universal line bundle over complex projective space $\mathbb{P}_1$. The result is simple because the metric is invariant under a circle action; the polynomial f depends only upon the absolute values of z and w.

Finally we close this section by proving Theorem 3.6. In **[D3]** this result was proved under an unnecessary assumption on the defining polynomial h. We have the following complete result.

Theorem 3.6. *Suppose that M is the zero set of a bihomogeneous polynomial. Suppose that f is a bihomogeneous polynomial that is positive on $M - \{0\}$. Then f agrees with the squared norm of a holomorphic polynomial mapping on M.*

Proof. By considering the square of h if necessary, we may assume that M is the zero set of a nonnegative bihomogeneous polynomial h. Let C be a positive number, to be chosen momentarily. By choosing appropriate integers a and b, we may assume that the function F_C defined by (37)

$$F_C(z, \bar{z}) = \|z\|^{2a} f(z, \bar{z}) + C\|z\|^{2b} h(z, \bar{z}) \tag{37}$$

is a bihomogeneous polynomial. Suppose we can choose C so that F_C is positive on the unit sphere. By Theorem 3.1 it follows that there is an integer d and a holomorphic polynomial mapping g so that

$$\|z\|^{2d} F_C(z,\bar{z}) = \|g(z)\|^2. \tag{38}$$

On M, we have $h = 0$. Thus (37) and (38) yield, for $z \in M$,

$$\|z\|^{2b+2d} f(z,\bar{z}) = \|g(z)\|^2 \tag{39}$$

which gives the conclusion of the theorem.

It remains to show that we can choose C sufficiently large to make F_C positive on the unit sphere S. Let η be the minimum value assumed by f on S. Note that η could be negative. Since M is a closed set, and S is compact, $M \cap S$ is compact. Since f is positive there and continuous, its minimum value on $M \cap S$, which we denote by ϵ, is positive. Since f is continuous, there is a $\delta > 0$ so that $f \geq \frac{\epsilon}{2}$ at all points within a distance δ of $M \cap S$.

We claim that there is a $\gamma > 0$ so that, for $z \in S$, $h(z,\bar{z}) \leq \gamma$ implies $f(z,\bar{z}) \geq \frac{\epsilon}{2}$. If this were false, then it would fail for $\gamma = \frac{1}{n}$. Hence there would be a sequence of points on the sphere so that $h(z_n,\bar{z}_n) \leq \frac{1}{n}$ whereas $f(z_n,\bar{z}_n) \leq \frac{\epsilon}{2}$. Since the sphere is compact, there would be a convergent subsequence satisfying the same inequalities. Call the limit z. By continuity we would have $f(z,\bar{z}) \leq \frac{\epsilon}{2}$, but $h(z,\bar{z}) = 0$. Thus $z \in S \cap M$, so $f(z,\bar{z}) \geq \epsilon$. This contradiction proves that such a γ exists.

Now we choose C so that $\eta + C\gamma \geq \frac{\epsilon}{2}$. Recall that $0 \leq h$. On the sphere where $h \leq \gamma$, we have

$$F_C = f + Ch \geq f \geq \frac{\epsilon}{2}. \tag{40}$$

On the sphere where $h \geq \gamma$, we have

$$F_C = f + Ch \geq \eta + C\gamma \geq \frac{\epsilon}{2}. \tag{41}$$

We have proved that $F_C \geq \frac{\epsilon}{2}$ on S. This finishes the proof of the theorem. □

4. Squared norms and proper mappings between balls

We begin this section with some discussion about proper mappings between domains in (perhaps different dimensional) complex Euclidean spaces. Let Ω and Ω' be bounded domains in $\mathbb{C}^n$ and $\mathbb{C}^N$. A holomorphic mapping $f : \Omega \to \Omega'$ is proper if $f^{-1}(K)$ is compact in Ω whenever K is compact in Ω'. In case such a mapping extends to be a continuous mapping of the boundaries, then such a map will be proper precisely when it maps the boundary $b\Omega$ to the boundary $b\Omega'$. This leads us to CR geometry. When these boundaries are smooth manifolds, and the holomorphic mapping f extends smoothly to the boundaries, we obtain a smooth CR mapping. Given two CR manifolds of hypersurface type it is generally impossible to find a nonconstant CR mapping between them. On the other hand, when the domain and targets are balls, there are many examples. Therefore we continue the discussion by studying proper holomorphic mappings between balls.

Let $f : B_n \to B_N$ be a proper holomorphic mapping. When f extends smoothly to the boundary we see that $\|z\|^2 = 1$ implies $\|f(z)\|^2 = 1$.

Observation 1. When $N < n$, every such map is a constant. This follows from the observation that positive dimensional complex analytic subvarieties of the ball are noncompact.

Observation 2. When $n = N = 1$, a proper map is necessarily a finite Blaschke product. Thus there are finitely many points a_j in B_1, positive integer multiplicities m_j, and an element $e^{i\theta}$ such that

$$f(z) = e^{i\theta} \prod_{j=1}^{K} \left(\frac{z - a_j}{1 - \overline{a}_j z}\right)^{m_j}. \tag{42}$$

Note that (42) shows that there is no restriction on the denominator. Every polynomial q that is not zero on the closed ball arises as the denominator of a rational function reduced to lowest terms.

When $n = N > 1$, it is a well known result of Pinchuk and Alexander that a proper holomorphic map between balls is necessarily an automorphism. In particular the map is a linear fractional transformation and the denominator is $1 - \langle z, a \rangle$.

In order to obtain analogues of finite Blaschke products, one must replace multiplication by the tensor product. Since the tensor product of complex vector spaces of dimensions n and k is of dimension nk, the target dimension increases. It is thus natural to consider proper maps from a given B_n to B_N for all N.

Observation 3. It is known that there are proper holomorphic mappings $f : B_n \to B_{n+1}$ that are continuous up to the boundary, but that are not smooth. It seems hopeless to attempt to classify such mappings. Forstneric **[F]** proved a beautiful result in the smooth case. If $n \geq 2$, and a proper map $f : B_n \to B_N$ is $N - n + 1$ times continuously differentiable at the boundary, then f must be a rational mapping. Later work by Cima–Suffridge [CS] showed that the denominator cannot vanish on the closed ball.

Forstneric's result suggests classifying the rational mappings. Many of the results in this paper arose from studying this problem. Consider for example the following result. **[D3]**.

Theorem 4.1. *Let $q : \mathbb{C}^n \to \mathbb{C}$ be a holomorphic polynomial, and suppose that q does not vanish on the closed unit ball. Then there is an integer N and a holomorphic polynomial mapping $p : \mathbb{C}^n \to \mathbb{C}^N$ such that*

1. *$\frac{p}{q}$ is a rational proper mapping between B_n and B_N;*

2. *$\frac{p}{q}$ is reduced to lowest terms.*

Remark 4.2. We want $\frac{p}{q}$ to be in lowest terms, or else we have the trivial example where $p(z) = q(z)(z_1, \ldots, z_n)$.

Proof. The result is trivial when q is a constant. It is easy when $n = 1$. When the degree d of q is positive, we simply define p by $p(z) = z^d \overline{q}(\frac{1}{\overline{z}})$ and easily see that this works. Such a proof cannot work in higher dimensions! The minimum integer N can be arbitrarily large even when $n = 2$ and the degree of q is also two.

Now assume $n \geq 2$. Suppose that $q(z) \neq 0$ on the closed ball. Let g be an arbitrary polynomial such that q and g have no common factor. Then there is a constant c so that

$$|q(z)|^2 - |cg(z)|^2 > 0 \tag{43}$$

for $\|z\|^2 = 1$. We set $p_1 = cg$.

By Theorem 4.3 below a polynomial $f(z, \overline{z})$ that is positive on the unit sphere agrees with a squared norm of a holomorphic polynomial mapping on the sphere. Therefore there is an integer N and polynomials $p_2, \ldots, p_N$ such that

$$|q(z)|^2 - |p_1(z)|^2 = \sum_{j=2}^{N} |p_j(z)|^2 \tag{44}$$

on the sphere. It then follows that $\frac{p}{q}$ does the job. □

It remains to state Theorem 4.3; its proof appears in **[CD1]**.

Theorem 4.3 (Catlin–D'Angelo). *Suppose that $f(z, \overline{z})$ is a (not necessarily homogeneous) real-valued polynomial on $\mathbb{C}^n$ and that $f(z, \overline{z}) > 0$ on the unit sphere. Then there is an integer N and a (holomorphic) polynomial mapping $g : \mathbb{C}^n \to \mathbb{C}^N$ such that $f(z, \overline{z}) = \|g(z)\|^2$ on the unit sphere.*

Theorem 4.3 thus states that, if (a strong form of) 1) holds on the sphere, then 3) holds on the sphere. Its proof proceeds in 2 steps. The first step is to add a variable, homogenize, and reduce to the bihomogeneous case. The second step uses the stronger statement in the bihomogeneous case provided by Theorem 3.1.

There are related theorems due to Lempert (**[L1]**, **[L2]**) and to Løw **[L]**. They state that positive functions on the boundaries of strongly pseudoconvex domains agree with squared norms of holomorphic mappings there. In Lempert's work, the boundary is real-analytic, the given positive function is real-analytic, and the holomorphic mapping takes values in an infinite-dimensional space. In Løw's work, the domain has C^2 boundary, the given positive function is continuous, the mapping takes values in a finite-dimensional space, but it is holomorphic on only the interior of the domain.

Theorem 4.3 can be used also to show that one can choose various components p of a proper holomorphic polynomial mapping arbitrarily, assuming only that they satisfy the necessary condition $\|p(z)\|^2 < 1$ on the sphere.

Theorem 4.4. *Suppose that p is a vector-valued polynomial on $\mathbb{C}^n$ and that $\|p(z)\|^2 < 1$ on the unit sphere. Then there is a polynomial mapping g such that $p \oplus g$ defines a proper holomorphic mapping between balls.*

Proof. Note that $1 - \|p(z)\|^2$ is a polynomial that is positive on the sphere. Hence we can find a holomorphic polynomial mapping g such that

$$1 - \|p(z)\|^2 = \|g(z)\|^2$$

on the sphere. We may assume that not both p and g are constant. Then $p \oplus g$ is a non-constant holomorphic polynomial mapping whose squared norm $\|p\|^2 + \|g\|^2$ equals unity on the sphere. By the maximum principle $p \oplus g$ is the required mapping. □

5. Proper mappings and invariance under finite groups

D'Angelo–Lichtblau **[DL]** solved the CR Spherical Space Form problem by proving Theorem 5.1 below. This has an interesting relationship to squared norms of holomorphic mappings.

Suppose that G is a finite group, and that $\Gamma = \pi(G)$ is a representation of G as a subgroup of the unitary group $U(n)$. We call Γ a finite unitary group, and we say that it is fixed point free if the only element in Γ with an eigenvalue of unity is the identity. We say that a holomorphic mapping f is Γ-invariant if $f \circ \gamma = f$ for all $\gamma \in \Gamma$.

Theorem 5.1 (**[DL]**). *Suppose that Γ is a fixed-point free unitary representation of a finite group. Suppose that $f = \frac{g}{h}$ is a (holomorphic) rational function invariant under Γ, and furthermore suppose that there is an integer N such that $f : S^{2n-1} \to S^{2N-1}$. In order that f be non-constant, Γ must be cyclic, and represented in one of the following three ways.*

(0.1) *Γ is the cyclic group of order m generated by ϵI. Here ϵ is a primitive m-th root of unity.*

(0.2) *Γ is the cyclic group of order $2r+1$ generated by $\epsilon I_k \oplus \epsilon^2 I_{n-k}$. Here ϵ is a primitive $(2r+1)$-st root of unity, and I_k denotes the k-th order identity matrix.*

(0.3) *Γ is the cyclic group of order seven generated by $\epsilon I_k \oplus \epsilon^2 \oplus I_j \oplus \epsilon^4 I_{n-k-j}$. Here ϵ is a primitive seventh root of unity.*

There are explicit formulas for the minimal monomial example in each of these cases. Both the proof of Theorem 5.1 and the formulas for the maps that do arise use properties of the Γ-invariant real-valued polynomial Φ defined below in (45). There are also interesting combinatorial aspects. In the first case the multinomial coefficients arise as squared absolute values of coefficients of the monomials; in the second case a new triangle of integers arises in this way. See **[D2]** and Chapter 5 in **[D1]** for combinatorial aspects of invariant holomorphic mappings.

Corollary 5.2. *Let Γ be a fixed-point free unitary representation of a finite group. Define its invariant polynomial Φ by*

$$\Phi_\Gamma(z, \bar{z}) = 1 - \prod_{\gamma \in \Gamma}(1 - \langle \gamma z, z \rangle). \tag{45}$$

Then Φ is a squared norm of a holomorphic mapping only when Γ is one of the three cases from Theorem 5.1. *In each case the holomorphic mapping is a proper polynomial mapping between balls that is Γ-invariant. Thus* 3) *holds only in these cases.*

6. Isometric imbedding for holomorphic bundles

Finally we reinterpret Theorems 2.13 and 3.1. Let f be a bihomogeneous polynomial that is positive away from the origin in $\mathbb{C}^{N+1}$. The link to bundles arises by first considering complex projective space $\mathbb{P}_N$, the collection of lines through the origin in $\mathbb{C}^{N+1}$. We have the usual open covering given by open sets U_j where $z_j \neq 0$. In U_j we define f_j by

$$f_j(z, \bar{z}) = \frac{f(z, \bar{z})}{|z_j|^{2m}}. \tag{46}$$

On the overlap $U_j \cap U_k$ these functions then transform via

$$f_k = \left|\left(\frac{z_j}{z_k}\right)^m\right|^2 f_j. \tag{47}$$

Since $\left(\frac{z_j}{z_k}\right)^m$ are the transition functions for the m-th power of the universal line bundle $\boldsymbol{U}^m$, the functions f_j determines an Hermitian metric on $\boldsymbol{U}^m$, called here a *special* metric.

The main purposes of this section are to reformulate Theorem 3.1 in this language and then to generalize it. First we reinterpret Theorem 2.13.

Suppose that f is a bihomogeneous polynomial of degree $2m$. Then it defines via (46) a metric on $\boldsymbol{U}^m$ if and only if it is positive as a function away from the origin. This is a sharp form of 1). This metric is already a pullback of the Euclidean metric if and only if 3) or 4) holds. Some tensor power of the bundle with itself is a pullback if and only if 5) holds. The Cauchy–Schwarz inequality implies plurisubharmonicity, which in turn is equivalent to the negativity of the curvature of the bundle, or the pseudoconvexity of the unit ball in the total space of the bundle.

The previous paragraph applies in particular to the function r_a from Proposition 2.10. When $a < 16$, this bihomogeneous polynomial is strictly positive away from the origin, and hence defines a metric on U^4 on $\mathbb{P}_1$. By varying the parameter a we see that the various positivity properties of bundle metrics are also distinct.

We next restate Theorem 3.1.

Theorem 6.1. *Let* (U^m, f) *denote the* m*-th power of the universal line bundle over* $\mathbb{P}_n$ *with special metric defined by* f. *Then there is an integer* d *so that* $(U^{m+d}, \|z\|^{2d} f(z, \bar{z}))$ *is a (holomorphic) pullback* $g^*(U, \|L(\zeta)\|^2)$ *of the standard metric on the universal bundle over* $\mathbb{P}_N$. *The mapping* $g : \mathbb{P}_N \to \mathbb{P}_N$ *is a holomorphic (polynomial) embedding and* L *is an invertible linear mapping.*

$$(U^m, f) \otimes (U^d, \|z\|^{2d}) = (U^{m+d}, \|z\|^{2d} f(z, \bar{z})) = (U^{m+d}, \|g(z)\|^2)$$

We have the bundles and metrics

$$\pi_1 : (U^m, f) \to \mathbb{P}_n$$

$$\pi_2 : (U^{m+d}, \|z\|^{2d} f) \to \mathbb{P}_n$$

$$\pi_3 : (U, \|L(\zeta)\|^2) \to \mathbb{P}_n.$$

Thus π_1 *is not an isometric pullback of* π_3, *but, for sufficiently large* d, π_2 *is such a pullback.*

This formulation suggests a generalization to more general Hermitian bundles. See **[CD3]** for the precise definitions of globalizable metric and the inequality (SGCS). The inequality (SGCS) is a sharp form of 6). A version of it arises also in Calabi's famous paper **[CL]** on isometric imbeddings of the tangent bundle. The main result of **[CD3]** is the following isometric imbedding theorem for holomorphic bundles.

Theorem 6.2. *Suppose that* M *is a compact complex manifold. Let* E *be a vector bundle of rank* p *over* M *with globalizable Hermitian metric* G. *Let* L *be a line bundle over* M *with globalizable Hermitian metric* R, *and suppose that* L *is negative and that* R *satisfies* (SGCS). *Then there is an integer* d_0 *so that, for all* d *with* $d \geq d_0$, *there is a holomorphic imbedding* h_d *with* $h_d : M \to \mathbf{G}_{p,N}$ *so that* $E \otimes L^d = h_d^* U_{p,N}$ *and* $GR^d = h_d^*(g_0)$.

The special case where the base manifold is complex projective space $\mathbb{P}_{n-1}$ gives us Theorem 3.1. We let E be a power U^m of the universal bundle, with metric determined by the bihomogeneous polynomial p, and we let L be the universal bundle U with the Euclidean metric. We also have a matrix analogue of Theorem 3.1, where we now let E be the bundle of rank r given by r copies of U^m. See **[D3]** for details.

Corollary 6.3. *Suppose that* $M(z, \bar{z})$ *is a matrix of bihomogeneous polynomials of the same degree that is positive-definite away from the origin. Suppose* R *is a bihomogeneous polynomial that is positive away from the origin and satisfies* (SGCS).

Assume also that $\{R < 1\}$ *is a strongly pseudoconvex domain. Then there is an integer* d *and a matrix* A *of holomorphic homogeneous polynomials such that*

$$R(z, \overline{z})^d M(z, \overline{z}) = A(z)^* A(z). \tag{48}$$

In particular we can choose $R(z, \overline{z}) = \|z\|^2$.

A matrix is C positive definite if and only if its components c_{ij} can be expressed as inner products $\langle e_i, e_j \rangle$ of basis elements, so we can factorize a positive definite matrix of constants in the form $C = A^*A$. When the entries depend real-analytically on parameters it is generally impossible to make A depend holomorphically on these parameters. Writing an operator-valued real-analytic function as in the right side of (48) is called *holomorphic factorization.* See **[RR]** for classical results about holomorphic factorization of operator-valued holomorphic functions of one complex variable. In the situation of Corollary 6.3 in this paper, one cannot factor M holomorphically, but one can factor $R^d M$ holomorphically when d is sufficiently large.

References

[CL] E. Calabi, Isometric imbedding of complex manifolds, Ann. of Math. 58 (1953), 1–23.

[CD1] David W. Catlin and John P. D'Angelo, A stabilization theorem for Hermitian forms and applications to holomorphic mappings, Math. Res. Lett. 3 (1996), 149–166.

[CD2] David W. Catlin and John P. D'Angelo, Positivity conditions for bihomogeneous polynomials, Math. Res. Lett. 4 (1997), 1–13.

[CD3] David W. Catlin and John P. D'Angelo, An isometric imbedding theorem for holomorphic bundles, Math. Res. Lett. 6 (1999), 1–18.

[CS] J. A. Cima and T. J. Suffridge, Boundary behavior of rational proper maps, Duke Math J. 35 (1988), 83–90.

[D1] John P. D'Angelo, Several Complex Variables and the Geometry of Real Hypersurfaces, CRC Press, Boca Raton, FL, 1992.

[D2] John P. D'Angelo, Invariant holomorphic maps, J. Geom. Anal. 6 (1996), 163–179.

[D3] John P. D'Angelo, Positivity conditions and squared norms of holomorphic polynomial mappings, in: Analysis, Geometry, Number Theory: The Mathematics of Leon Ehrenpreis (E. I. Grinberg et al., eds.), Contemp. Math. 251, Amer. Math. Soc., Providence, RI, 2000, 163–172.

[DL] John P. D'Angelo and Daniel A. Lichtblau, Spherical space forms, CR maps, and proper maps between balls, J. Geom. Anal. 2 (1992), 391–416.

[F] F. Forstneric, Extending proper holomorphic mappings of positive codimension, Invent. Math. 95 (1989), 31–62.

[HLP] G. H. Hardy, J. E. Littlewood, and G. Polya, Inequalities, At the University Press, Cambridge 1934.

[H] David E. Handelman, Positive Polynomials, Convex Integral Polytopes, and a Random Walk Problem, Lecture Notes in Math. 1282, Springer-Verlag, Berlin 1987.

[L1] L. Lempert, Imbedding Cauchy–Riemann manifolds into a sphere, Internat. J. Math. 1 (1990), 91–108.

[L2] L. Lempert, Imbedding pseudoconvex domains into a ball, Amer. J. Math. 104 (1982), 901–904.

[L] E. Løw, Embeddings and proper holomorphic maps of strictly pseudoconvex domains into polydiscs and balls, Math. Z. 190 (1985), 401–410.

[Q] Daniel G. Quillen, On the Representation of Hermitian Forms as Sums of Squares, Invent. Math. 5 (1968), 237–242.

[RR] M. Rosenblum and J. Rovnyak, The factorization problem for nonnegative operator valued functions, Bull. Amer. Math. Soc. 77 (1971), 287–318.

Department of Mathematics
University of Illinois
Urbana, IL 61801, U.S.A.
jpda@math.uiuc.edu

On a generalized reflection principle in $\mathbb{C}^2$

Peter Ebenfelt and Xiaojun Huang *

1. Introduction

The classical Schwarz reflection principle in the complex plane $\mathbb{C}$ can be formulated as follows. Assume that $F\ U \to \mathbb{C}$, where U is an open subset of $\mathbb{C}$, is a holomorphic mapping which extends continuously to a real-analytic curve M contained in the boundary of U. If F maps M into another real-analytic curve $M' \subset \mathbb{C}$, then F extends holomorphically to an open neighborhood of M. In higher dimensions the situation is much more complicated and the analogous statement fails in general as can be seen by easy examples. To find conditions under which a reflection principle holds is an interesting and challenging problem, which has been addressed by many mathematicians. We mention here only the papers **[BJT]**, **[BBR]**, **[BR1]**, **[BR2]**, **[DF]**, **[BHR]**, **[Hu1]**, **[HMM]** and refer the interested reader to the book **[BE]** and the recent survey article **[Hu2]** for a more detailed account of the history. However, we would like to point out that the following conjecture is still open in dimensions higher than three. Let $M, M' \subset \mathbb{C}^{n+1}$ be real-analytic hypersurfaces through the origin and $U \subset \mathbb{C}^{n+1}$ an open set containing M in its boundary. Assume that neither M nor M' contains a nontrivial complex analytic variety through 0. Then any holomorphic mapping $F\ U \to \mathbb{C}^{n+1}$ which extends continuously to M, sending M into M' and 0 to 0 extends holomorphically to an open neighborhood of 0. The second author **[Hu1]** proved this conjecture in $\mathbb{C}^2$.

If M' contains a nontrivial complex analytic variety through 0, then a local reflection principle cannot hold at 0 unless additional conditions (which exclude the possibility that all of M is mapped into the variety) are imposed on the mapping f. In **[HMM]** a reflection principle was proved in a situation (in $\mathbb{C}^{n+1}$) where M and M' are allowed to contain a complex analytic hypersurface through 0 and F extends as a suitably nondegenerate C^1 (continuously differentiable) mapping of M to M'. In the present paper, we consider this situation in $\mathbb{C}^2$ and by making use of the more complete geometric picture of real hypersurfaces in $\mathbb{C}^2$, we are able to show that a

*P. Ebenfelt was supported in part by a Swedish NFR; X. Huang was supported in part by NSF-9970439 and the Rutgers Research Council

reflection principle holds for similarly nondegenerate F which merely extend as continuous mappings. The main novelty here is Lemma 2.4 below which establishes the continuity of the derivative of the mapping along a certain vector field. Another crucial result is Proposition 3.1, which was obtained in **[HMM]** with the additional assumption of Hölder continuity.

As a consequence, by using the result in **[Hu1]**, we obtain the following; recall that a connected real-analytic hypersurface in $\mathbb{C}^{n+1}$ is called *Levi flat* if it is locally biholomorphic, near every point, to a real hyperplane, i.e., a $2n+1$ dimensional linear subspace of $\mathbb{R}^{2n+1} \cong \mathbb{C}^{n+1}$.

Theorem 1. *Let $M, M' \subset \mathbb{C}^2$ be real-analytic Levi non-flat hypersurfaces and $D \subset \mathbb{C}^2$ a domain (connected open set) with M in its boundary. Assume that $F\ D \to \mathbb{C}^2$ is a holomorphic mapping which extends continuously to M and sends M into M'. If*

(i) *$F(D)$ is not contained in a 1-dimensional complex analytic variety, and*

(ii) *F restricted to any nontrivial complex analytic variety contained in M is not constant,*

then F extends holomorphically to an open neighborhood of M.

As a corollary, we obtain the following result.

Corollary 1. *Let $M, M' \subset \mathbb{C}^2$ be real-analytic Levi non-flat hypersurfaces and $D \subset \mathbb{C}^2$ a domain with M in its boundary. Assume that $F\ D \to \mathbb{C}^2$ is a holomorphic mapping which extends continuously to M. If $F\ M \to M'$ is a finite-to-one mapping, then F extends holomorphically to an open neighborhood of M.*

Theorem 1 will be a consequence of the result in **[Hu1]** and the following, which is the main contribution of this paper.

Theorem 2. *Let $M \subset \mathbb{C}^2$ be a real-analytic Levi nonflat hypersurface which contains a nontrivial complex analytic variety E through 0. Let $D \subset \mathbb{C}^2$ be an open connected set which contains M in its boundary. Assume that $F\ D \to \mathbb{C}^2$ is a holomorphic mapping which extends continuously to M and sends M into another real-analytic hypersurface $M' \subset \mathbb{C}^2$. If*

(i) *$F(D)$ is not contained in a 1-dimensional complex analytic variety, and*

(ii) *$F|_E$ is not constant,*

then F extends holomorphically to an open neighborhood of 0 in $\mathbb{C}^2$.

Remark. As will become apparent from the proof of Theorem 2, condition (i) is in fact equivalent to $J_F \not\equiv 0$ in D, where J_F denotes the Jacobian determinant of F i.e., $J_F := \det(\partial F/\partial Z)$. Clearly, (i) is implied by the latter. The converse is a consequence of the fact that $F(M) \subset M'$. (Indeed, (i) does not imply $J_F \not\equiv 0$ in D without this condition.) We also mention that the presence of a complex analytic curve in M enables us to get the single-valued holomorphic extension directly. This is in contrast with the proof of the holomorphic extension in the finite type case (**[Hu1]**), where one has to split the proof in two separated steps. Step (a): get the multiple-valued extension (or get the correspondence extension in some literature), assuming the result in the following Step (b); Step (b): prove the single-valued property assuming the map already extends as a multiple-valued analytic maps. (This step is usually called the study of the regularity problem of correspondences (see **[Hu3]**).) Indeed, the proof of regularity for proper and CR holomorphic correspondences in Part II of **[Hu1]** (which was first circulated in its earlier version **[Hu3]**) is crucial for the construction of CR correspondence in Part I of **[Hu1]**. In passing, the second author points out that the remark to **[Hu3]** by Diederichs–Pinchuk on p. 1091 of **[DP]** apparently needs to be fundamentally corrected. The interested reader is referred to **[Hu1]** for the historical facts and related discussions. Here, the second author uses this opportunity to just emphasize the fact that in the (distributed lecture) notes mentioned in their remark, there was not even a single statement or sentence remotely related to the regularity problem of correspondences.

We would like to point out that if F in Theorem 2 is assumed to extend as a C^1 mapping to M, then the conclusion follows from the result in **[HMM]**. If F extends as a C^∞ mapping and J_F does not vanish to infinite order at 0 (which implies (i) in Theorem 2) then, in view of the reflection principle in **[BR2]**, condition (ii) in Theorem 2 is redundant. However, if only a finite degree of smoothness is required then condition (ii) is needed as the following example illustrates.

Example 1 (**[HMM]**). Let $M, M' \subset \mathbb{C}^2$ be given by

$$M = \{(z, w) \in \mathbb{C}^2 : \operatorname{Im} w = (\operatorname{Re} w)^2 |w|^{2k^2+k} |z|^{2k}\},$$
$$M' = \{(z, w) \in \mathbb{C}^2 : \operatorname{Im} w = (\operatorname{Re} w)^2 |z|^{2k}\}.$$

Consider

$$F(z, w) = (z w^{k+1/2}, w),$$

where k is even. Then F is a holomorphic mapping $D \to \mathbb{C}^2$, where

$$D = \{(z, w \ \operatorname{Im} w > (\operatorname{Re} w)^2 |w|^{2k^2+k} |z|^{2k}\},$$

which extends as a $C^{k,1/2}$ mapping $M \to M'$ but not as a holomorphic mapping in any neighborhood of 0 in $\mathbb{C}^2$. Observe that J_F vanishes to finite order at 0 and, hence, F satisfies condition (i) of Theorem 2. However, F does not satisfy condition (ii), since $F^{-1}(0)$ is the complex hyperplane $\{w = 0\} \subset M$.

We would like to conclude the introduction with an open problem. Recall that a real hypersurface $M \subset \mathbb{C}^{n+1}$ inherits a CR structure $T^{0,1}M \subset \mathbb{C}TM$ from the ambient space; the subspace $T_p^{0,1}M$ is the n-dimensional complex vector space consisting of those $X \in T^{0,1}p\mathbb{C}^{n+1}$ which are also in tangent to M. Sections of $T^{0,1}M$ are called CR vector fields. A locally integrable function on M is called a CR function if it is annihilated by all CR vector fields, and a mapping $F\ M \to M' \subset \mathbb{C}^{n+1}$ is called CR if its components F_j are CR functions. (The reader is referred to **[BE]** for an introduction to CR geometry and basic results.) An important result is the following. All continuous CR functions defined in a neighborhood of 0 in M arise as boundary values of holomorphic functions defined on one side of M if and only if M does not contain a complex analytic hypersurface though 0 (see **[Tr]**; see also **[BT1]** and **[Tu]**). Hence, if M and M' are as in Theorem 2, then the class of continuous CR mappings $F\ M \to M'$ is potentially larger than the class of boundary values of holomorpic mappings considered in Theorem 2 and one may ask: Does the conclusion of Theorem 2 also hold for the class of continuous CR mappings (which would then imply that the two classes coincide)? The answer is negative as is demonstrated by the following example.

Example 2. Let $M, M' \subset \mathbb{C}^2$ be given by

$$\begin{aligned} M &= \{(z, w) \in \mathbb{C}^2\ \operatorname{Im} w = (\operatorname{Re} w)|z|^2\} \\ M' &= \{(z, w) \in \mathbb{C}^2\ \operatorname{Im} w = (\operatorname{Re} w)h_k(z, \overline{z})\}. \end{aligned}$$

where h_k is defined as follows. Define the finite sequences $\{a_k\}$ and $\{b_k\}$ by

$$\phi_k(s,t) = \operatorname{Re}(s+it)^k = s^n + \sum_{j=0}^{k-1} a_j s^j t^{k-j}$$

$$\psi_k(s,t) = \operatorname{Im}(s+it)^k = \sum_{j=0}^{k-1} b_j s^j t^{k-j}$$

and set

$$h_k := \frac{\sum_{j=0}^{k-1} b_j |z|^{2(k-j)}}{1 + \sum_{j=0}^{k-1} a_j |z|^{2(k-j)}}.$$

Let $F = (z, g)$ with $g = -w^k$ for $s \leq 0$ and $g = w^k$ for $s > 0$. Then F is a finite to one C^{k-1}-differentiable CR mapping from M into M', which does not extend holomorphically to any neighborhood of 0. Observe that F does not extend holomorphically to any side of M, nor is it C^k-differentiable over M. Related to this example, we mention a result proved in **[BHR]** for algebraic CR manifolds, which, in particular, implies that if the degree of smoothness of the CR mapping F is sufficiently high relative to the degree of the target manifold M', then F does extend holomorphically (indeed, algebraically) to a neighborhood of 0.

Thus, the assumption that F in Theorem 2 extends holomorphically to one side of M cannot be removed. However, the authors have some hope that the following is true.

Conjecture. *Let $M, M' \subset \mathbb{C}^2$ be real-analytic Levi nonflat hypersurfaces. Then any C^∞-smooth CR (local) diffeomorphism $F\ M \to M'$ extends holomorphically to an open neighborhood of M in $\mathbb{C}^2$.*

Note that F in the conjecture, by known results mentioned above, extends holomorphically to an open neighborhood in $\mathbb{C}^2$ of a dense open subset of M, namely the set of points through which there are no complex analytic hypersurfaces contained in M. For future reference, we recall that a real hypersurface which does not contain a complex analytic hypersurface through a point $p \in M$ is said to be minimal at p. For a real-analytic hypersurface, minimality at a point is equivalent to being of finite type in the sense of Kohn and Bloom–Graham at this point. We also mention that if M and M' in addition are assumed to by algebraic in the conjecture, then the conclusion follows from the results in **[BHR]**.

2. Geometry of the Segre varieties near an infinite type point

For the proof of the Theorem 2, we will use some ideas in **[HMM]**. However there would be a serious difficulty if we try to directly modify it to the case of continuous mappings. This is because the argument in **[HMM]** needs to deal with the quotient of certain distributions coming from the differential of the map. In this paper, we will find a way to deal with this in the case of complex two-space by employing geometric information of the Segre invariant family near an infinite type point. The purpose of this section is to make a preparation towards such an effort.

Let M be a Levi non-flat real analytic hypersurface in $\mathbb{C}^2$ with $0 \in M$ an infinite type point. Then, after a holomorphic change of variables, we can assume that M near 0 is defined by an equation of the form: $t = s^k h(z, \overline{z}, s)$, where $k \in \mathbb{Z}^+$, $h(z, 0, s) = h(0, \overline{z}, s) \equiv 0$ and $h(z, \overline{z}, 0) \not\equiv 0$. Also, in what follows, we use (z, w), with $w = s + it$, for the coordinates of $\mathbb{C}^2$. Write $m(0)$ for the lowest degree of the non-vanishing terms in the Taylor expansion of $h(z, \overline{z}, 0)$ near 0. It is known (see **[Me]**) that k, $m(0)$ are holomorphic invariant of M. Moreover, in such a coordinates system, M contains the complex curve $E = \{w = 0\}$ and for each $p_0 = (z_0, 0)$ near 0 (which in what follows will be abbreviated $p_0 \approx 0$ M has the same k-value, but $m(p_0)$ may well depend on p_0.

Let $u(z, w) = t - s^k h(z, \overline{z}, s)$. Then the Levi determinant of M associated to the defining function u is given by

$$J_u := \det \begin{pmatrix} 0 & u'_z & u'_w \\ u'_{\overline{z}} & u''_{z\overline{z}} & u''_{w\overline{z}} \\ u'_{\overline{w}} & u''_{z\overline{w}} & u''_{w\overline{w}} \end{pmatrix}.$$

A direct computation shows that $J_u(z, \bar{z}, s) = s^k \Delta(z, \bar{z}, s)$ and the lowest degree of the non-vanishing terms in the Taylor expansion of $\Delta(z, \bar{z}, 0)$ near 0 is $m(0) - 2$. Let E_0 be the zero set of Δ in E, which is a thin set in E. Then it is clear that for each $p_0 \in E \setminus E_0$, $m(p_0) = 2$. Hence, we have

Lemma 2.1. *For each $p_0 \in E \setminus E_0$, there are holomorphic coordinates (z, w), vanishing at p_0, in which M can be defined, near $p_0 = (0, 0)$, by an equation of the form $t = s^k h(z, \bar{z}, s)$, where $h(z, \bar{z}, 0) = \pm|z|^2 + o(|z|^2)$ and $h(z, 0, s) = h(0, \bar{z}, s) \equiv 0$.*

Now assume that 0 is an infinite type point of M with $m(0) = 2$. We need some basic property of Segre varieties near such a point. To this aim, we set up the following notation. The reader is referred to **[Hu1]** for a more detailed explanation of the notation and definitions.

Choose two sufficiently small balls $\mathcal{P}, \widetilde{\mathcal{P}}$ centered at 0 with $\mathcal{P} \subset\subset \widetilde{\mathcal{P}}$. Making the radius of $\mathcal{P}$ much smaller than that of $\widetilde{\mathcal{P}}$, we can assume that for each $p \in \mathcal{P}$, the Segre variety Q_p of M associated to p is a closed complex hypersurface in $\widetilde{\mathcal{P}}$. Also, there is a side-reversing (with respect to M) diffeomorphic map from $\mathcal{P}$ into its image which is contained in $\widetilde{\mathcal{P}}$, called the conjugate map $\mathcal{R}$, such that $\mathcal{R}(p) \in Q_p$ for each $p \in \mathcal{P}$. Also, $\mathcal{R}|_M = \mathrm{id}$, $\mathcal{R}(p)$ and p always stay on different sides of M for $p \notin M$. For the system chosen above, we say that a point $p \in \mathcal{P}$ stays in the positive side of M if the connected component of $Q_p \setminus M$ containing $\mathcal{R}(p)$ is compact in $\widetilde{\mathcal{P}}$. In this case, we denote such a component by Q_p^c.

For each $p = (z, w) \in M \cap \mathcal{P}$ and a positive number ϵ, write $N_{\epsilon,p}^{\pm} = \{(z, w) \pm (0, i\tau\epsilon) : 0 < \tau \leq 1\}$, respectively, and $N_{\epsilon,p} = N_{\epsilon,p}^{+} \cup N_{\epsilon,p}^{-} \cup \{p\}$.

Lemma 2.2. *Assume $\mathcal{P}$ is sufficiently small and δ is a sufficiently small positive number. For any $p_0 = (z_0, w_0) \in \mathcal{P} \cap M$ with $w_0 = s_0 + it_0$, consider the set $N_{\epsilon,p_0}^{\pm}$ with $\epsilon < \delta|s_0|^k$. Then any point in N_{ϵ,p_0}^{-} stays in the positive side of M in the following cases:* (I) *$h(z, \bar{z}, 0) = |z|^2 + o(|z|^2)$, $s_0 > 0$;* (II) *$h(z, \bar{z}, 0) = |z|^2 + o(|z|^2)$, $s_0 < 0$, and k is even;* (III) *$h(z, \bar{z}, 0) = -|z|^2 + o(|z|^2)$, $s_0 < 0$ and k is odd. In all other situations, N_{ϵ,p_0}^{+} must stay in the positive side of M.*

Proof. In the following we fix a small $\widetilde{\mathcal{P}}$ but reserve the right to shrink $\mathcal{P}$. We first assume that $h(z, \bar{z}, 0) = |z|^2 + o(|z|^2)$, and let $p_0 = (z_0, s_0 + it_0) \in \mathcal{P}$ with $s_0 > 0$. We will verify that any point in N_{ϵ,p_0}^{-} stays in the positive side of M in the sense defined above, whenever $\epsilon < \delta s_0^k$ with $\delta > 0$ a sufficiently small number (but independent of the particular choice of p_0). Let $p_\tau = (z_0, w_\tau) = (z_0, s_0 + it_0 - i\tau\epsilon) \in N_{\epsilon,p_0}^{-}$. Then

$$Q_{p_\tau} = \left\{(z, w) \in \widetilde{\mathcal{P}} : \frac{w - \overline{w_\tau}}{2i} = \left(\frac{w + \overline{w_\tau}}{2}\right)^k h\left(z, \overline{z_0}, \frac{w + \overline{w_\tau}}{2}\right)\right\}.$$

Recall the notation $u = -t + s^k h(z, \bar{z}, s)$. It is then clear that M is strongly pseudoconvex near p_τ towards the side defined by $u(z, w) < 0$. Now, by a fundamental observation of Webster (**[We2]** or Lemma 2.2 of **[Hu1]**), when $\tau \ll 1$, p_τ stays in the positive side of M and the diameter of $Q^c_{p_\tau} (\subset \{u < 0\})$ is about $\sqrt{\tau}$. Notice that $\mathcal{R}(p_\tau)$ depends continuously on $\tau \in (0, 1]$. Hence, for each τ, if we can show $u(z, w) > 0$ for $(z, w) \in Q_{p_\tau}$ when $\delta_1 > |z| > \delta_2$, where δ_j $(j = 1, 2)$ are small positive numbers which are independent of the choice of p_0, then it is clear that after making $\mathcal{P}$ sufficiently small, any point in N^-_{ϵ, p_0} stays in the positive side of M and $Q^c_{p_\tau}$ stays in the side of $u < 0$. Pick $(z, w) \in Q_{p_\tau}$. Then we have

$$w = \overline{w_\tau} + 2i \left(\frac{w + \overline{w_\tau}}{2} \right)^k h \left(z, \overline{z_0}, \frac{w + \overline{w_\tau}}{2} \right).$$

On the other hand, $|t_0| = s_0^k |h(z_0, \overline{z_0}, s_0)| \lesssim s_0^k |z_0|$. Hence, when $\epsilon < \delta s_0^k$ with δ a sufficiently small number, we get $\operatorname{Im} w = \epsilon\tau + R_1$ and $\operatorname{Re}(w) = s_0 + R_2$, where $|R_1| < C^* s_0^k |z_0|$ and $|R_2| < C^* s_0 |z_0|$ with C^* a fixed number depending only on M and the initial choice of $\widetilde{\mathcal{P}}$. Therefore, to make $u = -\operatorname{Im}(w) + (\operatorname{Re}(w))^k h(z, \overline{z}, s) > 0$, it suffices to make

$$h(z, \overline{z}, s) > \frac{\tau\epsilon + R_1}{(s_0 + R_2)^k}.$$

On the other hand, by shrinking $\mathcal{P}$, we can make $|z_0|$ small; and by making $|z| < \delta_1$ with δ_1 a small but fixed number, we can make $h(z, \overline{z}, s) > \frac{1}{2}|z|^2$. Hence, after shrinking $\mathcal{P}$ and δ if necessary, we can find another small positive number $\delta_2 (< \delta_1)$ which is independent of the choice of p_0, such that when $(z, w) \in Q_{p_\tau}$ with $\delta_1 > |z| > \delta_2$, then $u(z, w) > 0$. As mentioned before, this implies that the whole line $N^-_{p_\tau, \epsilon}$ stays in the positive side of M. The other cases can be verified similarly and the details are left to the reader. □

Note that for $p \in \mathcal{P}$, $|u(p)|$ is proportional to the distance of p to M. For a small $\delta^* > 0$, let

$$U_{\delta^*} = \{ p = (z, w) \in \mathcal{P} : \operatorname{dist}(p, M) \leq \delta^* (|\operatorname{Re}(w)|)^k \}.$$

Then $U_{\delta^*} \setminus M$ has four simply connected components. After shrinking δ^*, if necessary, we can find two of them staying in the positive side of M. Denote the union of such two components by $U^+_{\delta^*}$ and the union of the other two by $U^-_{\delta^*}$. It is clear by Lemma 2.2. that we can assume that $U^+_{\delta^*}$ has M near 0 as part of its boundary. Since $\mathcal{R}$ is a diffeomorphic map, by suitably choosing $(\delta \gg)\, \delta^{**} > \delta^*$, we can assume that $\bigcup_{p \in U^+_{\delta^{**}}} Q^c_p$ contains the closure of $U^-_{\delta^*}$. Also, for each $p \in U^-_{\delta^*}$ there is a continuous family of complex analytic disks attaching to M, shrinking down to some point in M, whose image contains p (see **[E]** Proposition 4.3). By applying the Baouendi–Treves approximation theorem **[BT2]** and the maximum principle of holomorphic functions, it follows that any CR function extends holomorphically to $U^-_{\delta^*}$. Moreover, for any $\xi \in U^-_{\delta^*}$, $|h(\xi)| \leq \max_{p \in M \cap \widetilde{\mathcal{P}}} |h(p)|$. In what follows, we will fix the above δ^*. We will use the notation $O(A)$ for a small neighborhood of A whose size may be different in different contexts. Also, we write $M^+ = M \cap \{s > 0\}$ and $M^- = M \cap \{s < 0\}$.

Lemma 2.3. *Let $M, M' \subset \mathbb{C}^2$ be Levi non-flat real analytic hypersurfaces, which are of infinite type at* 0 *and have the m-value two at* 0. *Let F be a continuous CR map from M into M' such that $F(0) = 0$. Assume that neither of $F(M^+)$ nor $F(M^-)$ is contained in a complex analytic variety of dimension* 1. *Then F extends to a bounded holomorphic function in $U_{\delta^*} \cap O(0)$.*

Proof. First, by shrinking $\widetilde{\mathcal{P}}$, we can choose a small neighborhood $\mathcal{P}'$ of 0 such that $F(\widetilde{\mathcal{P}} \cap M) \subset\subset \mathcal{P}'$. We claim that the nondegeneracy for F as defined in the lemma implies that $F(M \setminus E) \cap E' = \emptyset$, where E' is the complex hypersurface of infinite type points of M' near 0. Indeed, the decomposition of M into local CR orbits (see e.g. **[BE]**) near 0 is given by $M = M^+ \cup E \cup M^-$ and similarly for M'. Since F is a CR mapping, $F(M^+)$, $F(M^-)$ and $F(E)$ are all contained in local CR orbits of M'. Since E' is a local CR orbit in M' which is also a 1-dimensional complex variety, and neither $F(M^+)$ nor $F(M^-)$, by assumption, is contained a 1-dimensional complex variety, the claim follows. This proves the claim. By the result in **[Hu1]**, F extends holomorphically to a sufficiently small neighborhood of $M^+ \cup M^-$. Since the m-value for M' is 2 at 0, one can easily verify that if $\mathcal{P}'$ is sufficiently small, $Q'_{q_1} = Q'_{q_2}$ with $q_1, q_2 \in \mathcal{P}'$ only if $q_1 = q_2$. Here, we use Q'_q to denote the Segre variety of M' associated to q.

Now, pick $p_0 = (z_0, s_0 + it_0) \in M \cap \mathcal{P}$ ($s_0 \neq 0$) and $\epsilon < \delta |s_0|^k$. Here δ, δ^* are arranged so that $\delta^* \ll \delta$. Without loss of generality, let us assume that $N^+_{p_0,\epsilon}$ stays in the positive side of M. (Otherwise, $N^-_{p_0,\epsilon}$ will stay in the positive side of M.) We mention that by the proof of Lemma 2.2, any point in the component of $U_{\delta^*} \setminus M$, containing $N^+_{p_0,\epsilon}$, stays also in the positive side of M; also, F extends holomorphically to the component containing $N^-_{p_0,\epsilon}$.

We will show that F admits a holomorphic extension to a neighborhood of $N^+_{p_0,\epsilon}$, which is uniformly bounded with respect to p_0. For this purpose, we let

$$\mathcal{V}_{p_0,\epsilon} = \{(p, q) \in O(N^+_{p_0,\epsilon}) \times \mathcal{P}' : F(Q^c_p) \subset Q'_q\}.$$

Then $\mathcal{V}_{p_0,\epsilon}$ is a complex analytic variety of dimension two in $O(N^+_{p_0,\epsilon}) \times \mathcal{P}'$. Write π and π' for the natural projection from $\mathcal{V}_{p_0,\epsilon}$ to $O(N^+_{p_0,\epsilon})$ and $\mathcal{P}'$, respectively. Write $\mathcal{V}^c$ for the irreducible component, which contains a neighborhood of $(p_\tau, F(p_\tau))$ for $\tau \ll 1$. Then, the same argument as in **[Hu1**, Lemma 3.1**]** shows that $\pi(\mathcal{V}^c)$ covers $N^+_{p_0,\epsilon}$. By our choice of $\mathcal{P}'$, we may define a map near $N^+_{p_0,\epsilon}$ which assigns to each $p \in O(N^+_{p_0,\epsilon})$ the uniquely determined q with $q = \pi' \circ \pi^{-1}(p)$. Clearly, this map gives the holomorphic extension of F to a neighborhood of $N^+_{p_0,\epsilon}$. Moreover, as in the proof of Lemma 3.1 of **[Hu1]**, $\mathcal{R}'(q) = F(\xi)$ for some $\xi \in Q^c_p$. Here $\mathcal{R}'$ is the corresponding conjugate function of M'. As remarked before, we thus conclude from the maximum principle that $|q|$ is uniformly bounded (with respect to q_0). This completes the proof. □

Now, let M, M' and F be as in Lemma 2.3. For each $(z, w) \in M$, applying the Cauchy estimate to the holomorphic function $F(z, w+\eta)$ in η for $|\eta| < \delta^*(|\operatorname{Re}(w)|)^k$, we get

$$|F'_w(z,w)| \lesssim \frac{\max_{|\eta|=\delta^*|s|^k} |F(z,w+\eta)|}{|\operatorname{Re}(w)|^k} \lesssim \frac{1}{|\operatorname{Re}(w)|^k}.$$

Also, since $|u(z+\xi,w)| \lesssim |\xi| \cdot |\operatorname{Re}(w)|^k$, applying Lemma 2.3, we see that $\{F(z+\xi,w)\}$ forms a uniformly bounded holomorphic family in ξ for $|\xi| < \delta_0$ with δ_0 a fixed but sufficiently small number. Moreover $F(z+\xi,w)$ converges uniformly on compact sets to $F(\xi,0)$ as $(z,w) \to 0$. (Here $(z,w) \in M$ acts as a parameter.) Therefore, it follows that for any integer n, $\partial^n F/\partial z^n$ extends continuously from $(M \setminus E) \cap \mathcal{P}$ to $M \cap \mathcal{P}$. Set

$$L = \frac{\partial}{\partial z} - \frac{u'_z}{u'_w}\frac{\partial}{\partial w},$$

which is a $(1,0)$-vector field tangent to M. Hence, the restriction of L to M is a section of $T^{1,0}M$ and its complex conjugate is a basis for the CR vector fields on M. Since $u'_w = 1/2 + o(1)$ and $|u'_z| \lesssim |z||\operatorname{Re}(w)|^k$, the function $L(F) = F'_z - u'_z F'_w/u'_w$, which as mentioned above is real analytic over $M \setminus E$ in view of the result in **[Hu1]**, is bounded in $M \setminus E$ near 0 and $\lim_{p\to 0} L(F)(p) = F'_z(0)$, for $p \in M \setminus E$. Now, for p_0 sufficiently close to 0 such that $m(p_0) = m'(F(p_0)) = 2$, we can also choose normal coordinates (as given by Lemma 2.1) for M and M' near p_0 and $F(p_0)$, respectively. By repeating the same analysis as above in the new coordinates near p_0, we conclude that there exists a real-analytic nonvanishing $(1,0)$-vector field $\widetilde{L}$ tangent to M near p_0 such that $\widetilde{L}(F)$ has a limit at p_0. Since the dimension of $T^{1,0}M$ is one, $\widetilde{L} = a(z,w)L$, for some real-analytic nonvanishing function $a(z,w)$, we deduce that $L(F)$ has a limit at p_0 and $L(F)$ is continuous in a neighborhood of 0 in M. Summarizing the above, we obtain the following lemma, which is crucial for the proof of Theorem 2.

Lemma 2.4. *Let M, M' and F be as in Lemma 2.3. Then $L(F) = F'_z - u'_z F'_w/u'_w$ extends continuously from $M \setminus E$ to a neighborhood of 0 in M.*

3. Proof of the Theorem 2

With the aid of Lemma 2.4, we can now complete the proof of Theorem 2 as follows.

Let M, M' be of infinite type at 0, and let F be a continuous CR map from M into M' with $F(0) = 0$ which extends holomorphically to one side D of M such that F is continuous over $D \cup M$. Let $\mathcal{S} \subset M$ be a real analytic totally real submanifold of real dimension 2 passing through $p = 0$ such that $\mathcal{S} \cap E$ is a real analytic curve near 0. In what follows, we shall assume, without loss of generality, that E is still $\{w = 0\}$ but $\mathcal{S} = \mathbb{R}^2 \cap O(0)$. Moreover, we assume that the positive $\operatorname{Im} w$-axis is the outward normal direction of M along $\mathcal{S}$ near 0 with respect to the D-side. Observe that in these coordinates the defining equation for M is in general not of the form given by Lemma 2.1. For a small $\widetilde{\delta} > 0$, write $\mathcal{W}^{\pm} = \{(z, w) : \pm\widetilde{\delta} \operatorname{Im} w > |\operatorname{Im} z|\}$. Then $\mathcal{W}^{+} \subset O(0) \setminus D \cup M$ and $(\overline{z}, \overline{w}) \in D$ for $(z, w) \in \mathcal{W}^{+}$ near the origin. Let $\beta^{*}(\in \mathbb{R}) \approx 0$, $\beta = (\beta_1, \beta_2) \in \mathbb{R}^2$ with $\beta_2 \approx 1$ and $\beta_1 \approx 0$. Write $\Delta_\epsilon = \{\zeta \in \mathbb{C} : |\zeta| < \epsilon\}$, $\Delta_\epsilon{}^{+} = \{\zeta \in \Delta_\epsilon : \operatorname{Im} \zeta > 0\}$. For ϵ sufficiently small, we define ψ_{β,β^*} by $\zeta \longrightarrow (\beta_1\zeta + \beta^*, \beta_2\zeta)$. Then it is clear that $\psi_{\beta,\beta^*}(\Delta \setminus (\Delta_\epsilon^{+} \cup (-\epsilon, \epsilon)) \subset D$. In what follows, we will also write $(\operatorname{Re}(z), \operatorname{Re}(w)) = X$.

We choose the normal coordinates for M' so that it is defined near 0 by an equation of the form $t = s^{k'}\chi(z, \overline{z}, s)$ with $\chi(z, 0, s) = \chi(0, \overline{z}, s) = 0$ and $k' > 0$. Write $F = (f, g)$. We shall refer to g as the normal component of F. It is clear that $g(z, 0) = 0$ since $F(E) \subset E' = \{w = 0\}$. On the other hand, by condition (i) in Theorem 2 and the remark at the beginning of the proof of Lemma 2.3, $g \not\equiv 0$ for $(z, w) \in M \setminus E$ since $F(M \setminus E) \cap E' = \emptyset$. The following is another crucial lemma for the proof of Theorem 2.

Proposition 3.1. *There is a positive integer n and a point $p_0 = (\beta_0, 0) \in \mathcal{S} \cap E$ near 0 such that for any positive integer m, $1/\overline{g^m(X)}$, initially defined over $\mathcal{S} \setminus E$, has a unique holomorphic extension $h_m(z, w)$ to $\mathcal{W}^{+} \cap O(p_0)$. Moreover, $|h_m(z, w)| \lesssim 1/|\operatorname{Im} w|^N$ for some integer N depending only on m, and $h_m(z, w)$ has the limit $1/\overline{g^m(X)}$ as $(z, w) \to X \in \mathcal{S} \setminus E$.*

Proposition 3.1 readily follows from Proposition 3.2 of **[HMM]** when F is assumed to be Hölder continuous. (In this case, h_m was actually proved to be Hölder continuous over $\overline{\mathcal{W}^{+}} \cap O(p_0)$ **[HMM]**.) In the present context, the proof can be carried out by a similar argument. Let us start with the following Riemann–Hilbert problem. We shall use the notation $\Delta_r^{+} = \{z \in \mathbb{C} : |z| < r, \operatorname{Im}(z) > 0\}$, and $\Delta^{+} = \Delta_1^{+}$.

Lemma 3.2. *Let $\phi(z, \tau) = u(z, \tau) + iv(z, \tau)$ be a nonconstant holomorphic function of z in Δ^{+}, which is continuous over $\overline{\Delta^{+}} \times O(\tau_0)$, where $O(\tau_0)$ is a small neighborhood of τ_0 in the parameter τ-space. Suppose that for $x \in (-1, 1)$, $v(x, \tau) = u(x, \tau)P(x, \tau)$ with $P(x, \tau) \in C((-1, 1) \times O(\tau_0))$. Then there exists a non-negative integer n, a small $\epsilon_0 > 0$, and a positive constant C_0 such that $\phi(z, \tau) = z^n\phi^{*}(z, \tau)$, where $\phi^{*}(z, \tau)$ is a holomorphic function of z in $\Delta_{\epsilon_0}^{+}$ and*

$$\sup_{0 \le y < \epsilon_0} \int_{-\epsilon_0}^{\epsilon_0} |\phi^{*}(x + iy, \tau)| \, dx < C_0, \qquad \sup_{0 \le y < \epsilon_0} \int_{-\epsilon_0}^{\epsilon_0} \left| \frac{1}{\phi^{*}(x + iy, \tau)} \right| dx < C_0$$

for $\tau \in O(\tau_0^{*})$, where τ_0^{*} is some point close to τ_0.

Proof. Write $\phi^{+}(z, \tau) = \phi(z, \tau)$ and $\phi^{-}(z, \tau) = \overline{\phi(\overline{z}, \tau)}$ for $z \in \overline{\Delta^{-}} = \{z \in \mathbb{C} : |z| \le 1, \operatorname{Im}(z) < 0\}$. Then for $x \in (-1, 1)$, one has

$$v(x, \tau) = \frac{1}{2i}(\phi^{+}(x, \tau) - \phi^{-}(x, \tau)), \quad u(x, \tau) = \frac{1}{2}(\phi^{+}(x, \tau) + \phi^{-}(x, \tau)).$$

Substituting this into the equation $v = uP(x,\tau)$, we obtain

$$\frac{1+iP(x,\tau)}{1-iP(x,\tau)}\phi^-(x,\tau) = \phi^+(x,\tau).$$

Now, let ϵ_0 be a sufficiently small positive integer and $\eta(x)$ a smooth cutoff function supported in $(-3\epsilon_0, 3\epsilon_0)$ with $\eta(x) \equiv 1$ for $|x| < 2\epsilon_0 \ll 1/2$. Consider the following Cauchy integral

$$\Gamma(z,\tau) = \frac{1}{2\pi i}\int_{-\infty}^{+\infty}\left(\eta(\xi)\log\frac{1+iP(\xi,\tau)}{1-iP(\xi,\tau)}\right)\frac{d\xi}{\xi - z}.$$

By the Plemelj formula, we have

$$e^{\Gamma^+(x,\tau)-\Gamma^-(x,\tau)} = \left(\frac{1+iP(x,\tau)}{1-iP(x,\tau)}\right)^{\eta(x)}$$

for almost every $x \in (-1,1)$. Here for $z \notin (-1,1)$, $\Gamma^{\pm}(z,\tau)$ is defined to be the value of $\Gamma(z,\tau)$ for $z \in \Delta^{\pm}$, respectively; and the value $\Gamma^{\pm}(x,\tau)$ on $x \in (-1,1)$ is understood as the non-tangential boundary limit of $\Gamma^{\pm}(z,\tau)$ as $z(\in \Delta^{\pm}) \to x$, respectively. Moreover, if we denote by $\mathcal{H}_r$ the Hilbert transform on $\mathbb{R}$ defined by $\mathcal{H}_r(u)(x) = \frac{1}{\pi}\int_{-\infty}^{+\infty}\frac{u(t)}{t-x}\,dt$, then

$$\Gamma^{\pm}(x,\tau) = \frac{1}{2}\left(\pm u_\tau(x) - i\mathcal{H}_r(u_\tau)(x)\right),$$

where

$$u_\tau(\xi) = \eta(\xi)\log\frac{1+iP(\xi,\tau)}{1-iP(\xi,\tau)}.$$

Recall that on the unit circle, the Hilbert transform $\mathcal{H}$, which maps u to its harmonic conjugate vanishing at 0, and the Cauchy singular integral operator $\mathcal{S}$ are related by the formula

$$\mathcal{H}(\phi) = i\mathcal{S}(\phi) - \frac{1}{2\pi}\int_{|\xi|=1}\frac{\phi(\xi)}{\xi}\,d\xi.$$

Also, by a classical result on the Hilbert transform on the unit circle (see **[Z]**, Theorem 2.11, p. 254, and its proof on p. 257 of **[Z]**),

$$\int_0^{2\pi} e^{|\mathcal{H}(\phi)|}\,d\theta < \frac{4\pi}{\cos(1)},$$

for any continuous function ϕ with $|\phi| \le 1$. Hence, after making use of a linear fractional transformation and basic properties of H^1-spaces, we can find some fixed constant C^* such that for $\tau \approx \tau_0$

$$\sup_{0\le y<\epsilon_0}\int_{-\epsilon_0}^{\epsilon_0} e^{|\Gamma(x\pm iy,\tau)|}\,dx < C^*. \tag{3.1}$$

Since $\phi^+(x,\tau)e^{-\Gamma^+(x,\tau)} = \phi^-(x,\tau)e^{-\Gamma^-(x,\tau)}$, it follows that $\phi^+(x,\tau)e^{-\Gamma^+(x,\tau)}$ extends holomorphically in z to the disk Δ_{ϵ_0}. Denote its extension by $\widetilde{\phi}(z,\tau)$. Then (3.1) can also be used to show that $\{\widetilde{\phi}(z,\tau)\}$ is a normal family with parameter $\tau(\approx \tau_0)$ for $|z| < \epsilon_0$. Meanwhile, since $\phi^{\pm}(z,\tau)$ and $e^{\Gamma^{\pm}(z,\tau)}$ depend continuously on τ when $z \notin (-1,1)$, we conclude that $\lim_{\tau\to\tau^*} \widetilde{\phi}(z,\tau) = \widetilde{\phi}(z,\tau^*)$ for any $\tau^* \in O(\tau_0)$. Write $\widetilde{\phi}(z,\tau) = z^{k_\tau}\widetilde{\phi}^*(z,\tau)$ with $\widetilde{\phi}^*(0,\tau) \neq 0$. By the Hurwitz theorem, k_τ is upper semi-continuous in τ. Now, since k_τ only takes integer value, we can find a point τ_0^* sufficiently close to τ_0 and an integer n, such that for $\tau \approx \tau_0^*$, $k_\tau = n$ and $\widetilde{\phi}^*(\xi,\tau)$ has no zeros over $\Delta_{\epsilon_0} \setminus \{0\}$ (after shrinking ϵ_0 (if necessary). The desired $\phi^*(z,\tau)$ is obtained by $\phi^*(z,\tau) = \widetilde{\phi}^*(z,\tau)e^{\Gamma^+(z,t)}$. Moreover, by (3.1), we obtain

$$\sup_{0\le y<\epsilon_0} \int_{-\epsilon_0}^{\epsilon_0} |\phi^*(x+iy,\tau)|\,dx, \qquad \sup_{0\le y<\epsilon_0} \int_{-\epsilon_0}^{\epsilon_0} \frac{1}{|\phi^*(x+iy,\tau)|}\,dx < C_0$$

for some fixed constant C_0 independent of $\tau \approx \tau_0^*$. This completes the proof. □

Proof of Proposition 3.1. Assume, without loss of generality, that $g(\beta^*,t) \not\equiv 0$ for $\beta^*(\in \mathbb{R}) \approx 0$ and $t < 0$. Write $\tau = (\beta_1,\beta_2,\beta^*) \approx (0,1,0)$, and write $g_\tau = \overline{g \circ \psi_\tau(\overline{\zeta})}$, where we recall that $\psi_\tau(\zeta) = (\beta_1\zeta + \beta^*, \beta_2\zeta)$. Then $g_\tau(\zeta)$ depends continuously on $\tau = (\beta_1,\beta_2,\beta^*)$ and $\zeta \in (-\epsilon,\epsilon)$. Also, it extends holomorphically to Δ_ϵ^+ in ζ. Note that for $\zeta \in (-\epsilon,\epsilon)$, $\psi_\tau(\zeta) \in M$ and, hence, $\mathrm{Im}(g_\tau(\zeta)) = -(\mathrm{Re}(g_\tau(\zeta)))^m\chi(f\circ\psi_\tau, \overline{f\circ\psi_\tau}, \mathrm{Re}(g_\tau(\zeta))$. Applying Lemma 3.2, we obtain the existence of a parameter $\tau_0 = (\beta_1^0, \beta_2^0, \beta_0^*)$ and an integer n such that for the sufficiently small $\epsilon_0 < \epsilon$ and $\tau \approx \tau_0$, $g_\tau(\zeta) = \zeta^n g_\tau^*(\zeta)$ with g_τ^* holomorphic in $\Delta_{\epsilon_0}^+$ and

$$\sup_{0\le y\le\epsilon_0} \int_{-\epsilon_0}^{\epsilon_0} |g_\tau^*(x+iy)|\,dx, \qquad \sup_{0<y\le\epsilon_0} \int_{-\epsilon_0}^{\epsilon_0} \frac{1}{|g_\tau^*(x+iy)|}\,dx$$

uniformly bounded with respect to $\tau(\approx \tau_0)$. Hence, the Cauchy formula

$$1/g_\tau^*(\zeta) = \frac{1}{2\pi i}\int_{\partial\Delta_{\epsilon_0}^+} \frac{1}{g_\tau^*(z)(z-\zeta)}\,dz, \quad \zeta \in \Delta_{\epsilon_0}^+,$$

and the above estimates imply that for $\zeta \in \Delta^+ \cap O(0)$

$$\frac{1}{|g_\tau^*(\zeta)|} \lesssim \frac{1}{|\mathrm{Im}(\zeta)|}$$

uniformly in $\tau(\approx \tau_0)$.

Denote by z_0 the point $(\beta_0^*, 0)$. Shrinking $\widetilde{\delta}$ in the definition of $\mathcal{W}^+$ if necessary, then the union of the image of the disks $(z,w) = \psi_\tau(\zeta) = (\beta_1\zeta + \beta^*, \beta_2\zeta)$, for $\tau \approx \tau_0$, $\zeta \in \{\zeta, \mathrm{Im}(\zeta) > 0\} \cap O(0)$ fills the set $\mathcal{W}^+ \cap O(z_0)$. Hence, by setting $(z,w) = \psi_\tau(\overline{\zeta})$ and using $\mathrm{Im}\, w = \beta_2 \mathrm{Im}\,\zeta$,

$$h_1(z,w) = \frac{1}{\overline{g((\overline{z},\overline{w}))}} = \frac{1}{|\zeta|^n|g_\tau^*(\zeta)|} \lesssim \frac{1}{|\mathrm{Im}\, w|^{n+1}}$$

for $(z, w) \in \mathcal{W}^+ \cap O(z_0)$. Since $g \not\equiv 0$, we have $F(M \setminus E) \subset M' \setminus E'$ as already pointed out in the proof of Lemma 2.3. Thus $F \in C^\omega(M \setminus E)$ by **[Hu1]**. The proof of Proposition 3.1 now follows. □

Proof of Theorem 2. The fact that the 1-dimensional complex analytic variety E through 0 is contained in the real hypersurface $M \subset \mathbb{C}^2$ implies that E is a complex (nonsingular) curve, as is well known, and the restriction of F to E is a holomorphic mapping $E \to M' \subset \mathbb{C}^2$. Since $F|_E$ is not constant, it is locally finite-to-one and the image $F(E) = E' \subset M'$ is a 1-dimensional complex analytic variety through 0. Hence, E' is also a complex (nonsingular) curve. To complete the proof of Theorem 2, in view of the Hanges–Treves propagation theorem **[HT]**, we need only show that F extends holomorphically to a neighborhood of some point in E. Thus, in what follows, we reserve the right to shift the origin in M to any nearby point $p \in E \subset M$. By applying Lemma 2.2, we may assume, without loss of generality, that $m(0) = m'(0) = 2$, $F(0) = 0$, and $F|_E$ is locally invertible near 0. Shifting again to a nearby point if necessary, we may also assume that the conclusion of Proposition 3.1 holds with $p_0 = 0$.

Let $u = t + o(|z, w|)$ be a real analytic defining function of M, and let $\mathcal{S}$ and its wedges $\mathcal{W}^\pm$ be chosen as before. Recall the $(1, 0)$-vector field

$$L = \frac{\partial}{\partial z} - \frac{u_z}{u_w}\frac{\partial}{\partial w},$$

which is tangent to M. We also use L for the restriction of this vector field to M. By Lemma 2.4, $L(F)$ is be continuous over M and $L(f)(0) = f'_z(0) \neq 0$. Also, recall that M' is defined by $t = s^k \chi(z, \bar{z}, s)$ with $\chi(z, \bar{z}, 0) = \pm|z|^2 + o(|z|)$, $\chi(z, 0, s) = 0$. Since $f(M) \subset M'$, we obtain $\mathrm{Im}(g(z)) = (\mathrm{Re}(g(z)))^m \chi(f(z), \overline{f(z)}, \mathrm{Re}(g(z)))$ for $z \in M$. By applying the implicit function theorem, we obtain the following functional equation

$$\overline{g(z)} = g(z) + (g(z))^m G(f(z), \overline{f(z)}, g(z)), \tag{3.2}$$

for $z \in M$, where $G(a, \bar{a}, 0)$ does not have any harmonic terms and $G(a, b, 0) = c_0 ab + o(ab)$ with $c_0 \neq 0$.

Now, applying $\overline{L}$ to (3.2), we obtain

$$\overline{Lg} = (g(z, w))^k \frac{\partial G}{\partial b}\big(f(z, w), \overline{f(z, w)}, g(z, w)\big)\overline{Lf}. \tag{3.3}$$

Since $\overline{L(F)}$ is continuous along $\mathcal{S}$, and $\overline{L(F)}$ extends holomorphically to the wedge $\mathcal{W}^+$ and blows up at most polynomially when approaching the edge $\mathcal{S}$, we conclude that $\overline{L(F)}$ has a unique holomorphic extension to $\mathcal{W}^+$, which is continuous up to $\mathcal{S}$ (see **[BJT]** or **[BE]** for instance). In particular, we see that $1/\overline{L(f)}$ has an extension which is holomorphic in $\mathcal{W}^+$ and continuous in $\overline{\mathcal{W}^+} \cap O(0)$. The reader should be warned that the holomorphic extension of $L(F)$ from $\mathcal{S}$ into $\mathcal{W}^-$ need not coincide with the $\mathbb{C}^2$-valued function obtained by applying the $(1, 0)$-vector field L to the holomorphic mapping F in $\mathcal{W}^-$.

Next, since

$$g(z,w) = \overline{g(z,w)} + (\overline{g(z,w)})^k \overline{G(f(z,w), \overline{f(z,w)}, g(z,w))}, \tag{3.4}$$

we obtain

$$\frac{\overline{Lg}}{\overline{L(f)}\overline{g}^k} = \left(1 + (\overline{g})^{k-1}\overline{G(f,\overline{f},g)}\right)^k \frac{\partial G}{\partial b}(f,\overline{f},g). \tag{3.5}$$

By the discussion above and Proposition 3.1, $\overline{Lg}/(\overline{L(f)}\overline{g^k})$ has a unique holomorphic extension $\chi(z,w)$ which is holomorphic in $\mathcal{W}^+ \cap O(0)$ and blows up at most polynomially when $(z,w) \to \mathcal{S}$. Hence, $\chi(z,w)$ has the limit, in the sense of distributions,

$$\left(1 + (\overline{g(z,w)})^{k-1}\overline{G(f(z), \overline{f(z)}, g(z))}\right)^k \frac{\partial G}{\partial b}(f(z,w), \overline{f(z,w)}, g(z,w)) \tag{3.6}$$

along $\mathcal{S}$ and, since (3.6) is continuous over $\mathcal{S}$, we conclude that $\chi(z,w) \in \mathrm{Hol}(\mathcal{W}^+) \cap C(\overline{\mathcal{W}^+} \cap O(0))$.

Since $\partial G^2/\partial a \partial b(0) \neq 0$, by applying the implicit function theorem to (3.5) in which we have substituted (3.4), we deduce that near $0 \in \mathcal{S}$,

$$f(X) = \sigma_1\big(\overline{f(X)}, \overline{g(X)}, \chi(X)\big)$$

where $X := \mathrm{Re}(z,w)$ and σ_1 is some function holomorphic in its argument. By the formula (3.4), we can find a σ_2 holomorphic in its argument such that for $X \in \mathcal{S}$ near 0, $g(X) = \sigma_2\big(\overline{f(X)}, \overline{g(X)}, \chi(X)\big)$.

Since $F(X) = (f(X), g(X))$ extends holomorphically to $\mathcal{W}^-$ and, for $j = 1, 2$, $\sigma_j\big(\overline{f(X)}, \overline{g(X)}, \chi\big)$ extend holomorphically to $\mathcal{W}^+$, it follows that F extends holomorphically to an open neighborhood of 0 by the classical edge of the wedge theorem, As mentioned above, by the Hanges–Treves theorem, F extends holomorphically to a neighborhood of E. This completes the proof of Theorem 2. □

Proof of Theorem 1. Let $p_0 \in M$ be a point of finite type, i.e., one through which M does not contain any nontrivial complex curves. By the same argument used in the beginning of the proof of Lemma 2.3 and condition (i), we conclude that $F(p_0)$ is of finite type in M. The holomorphic extension of F across p_0 follows from the result in **[Hu1]**. If $p_0 \in M$ is a point of infinite type, then the holomorphic extension across p_0 follows from Theorem 2. □

References

[BBR] M. S. Baouendi, S. Bell and L. P. Rothschild, Mappings of three-dimensional CR manifolds and their holomorphic extension, Duke Math. J. 56 (1988), 503–530.

[BE] M. S. Baouendi, P. Ebenfelt and L. P. Rothschild, Real submanifolds in complex spaces and their mappings, Princeton Math. Ser. 47, Princeton Univ. Press, 1999.

[BHR] M. S. Baouendi, X. Huang and L. P. Rothschild, Regularity of CR mappings between algebraic hypersurfaces, Invent. Math. 125 (1996), 13–36.

[BJT] M. S. Baouendi, H. Jacobowitz and F. Treves, On the analyticity of CR mappings, Ann. of Math. 122 (1985), 365–400.

[BR1] M. S. Baouendi and L.P. Rothschild, Germs of CR maps between real analytic hypersurfaces, Invent. Math. 93 (1988) 481–500.

[BR2] M. S. Baouendi and L. P. Rothschild, A general reflection principle in $\mathbb{C}^2$, J. Funct. Anal. 99 (1991), 409–442.

[BT1] M. S. Baouendi and F. Treves, About the holomorphic extension of CR functions on real hypersurfaces in complex space, Duke Math. J. 51 (1984), 77–107.

[BT2] M. S. Baouendi and F. Treves, A property of the functions and distributions annihilated by a locally integrable system of complex vector fields, Ann. of Math. 113 (1981), 387–421.

[DF] K. Diederich and J.-E. Fornaess, Proper holomorphic mappings between real-analytic pseudoconvex domains in $\mathbb{C}^n$, Math. Ann. 282 (1988), 681–700.

[DP] K. Diederich and S. Pinchuk, Proper holomorphic maps in dimension 2 extend, Indiana Univ. Math. J. 44 (1995), 1089–1126.

[E] P. Ebenfelt, On the unique continuation problem for CR mappings into nonminimal hypersurfaces, J. Geom. Anal. 6 (1996), 385–405.

[HT] N. Hanges and F. Treves, Propagation of holomorphic extendability of CR functions, Math. Ann. 263 (1983), 157–177.

[Hu1] X. Huang, Schwarz reflection principle in complex space of dimension two, Comm. Partial Differential Equations 21 (1996), 1781–1828.

[Hu2] —, On some problems in several complex variables and CR geometry, Proceedings of ICCM (S. T. Yau, ed.), AMS/IP Stud. Adv. Math. 20, 2001, 383–396.

[Hu3] X. Huang, Regularity of holomorphic correspondences and applications to the mapping problem, preprint, circulated in August 1995.

[HMM] X. Huang, J. Merker and F. Meylan, Mappings between degenerate real analytic hypersurfaces in $\mathbb{C}^n$, in: Analysis, Geometry, Number Theory: The Mathematics of Leon Ehrenpreis (E. I. Grinberg et al., eds.), Contemp. Math. 251, Amer. Math. Soc., Providence, RI, 2000, 321–338.

[Me] F. Meylan, A reflection principle in complex space for a class of hypersurfaces and mappings, Pacific J. Math. 169 (1995), 135–160.

[Tr] J. M. Trepreau, Sur le prolongement holomorphe des fonctions CR définies sur une hypersurface réelle de classe C^2 dans $\mathbb{C}^n$, Invent. Math. 83 (1986), 583–592.

[Tu] A. E. Tumanov, Extension of CR functions into a wdge from a manifold of finite type, Mat. Sb. 136 (1988), 128–139 = Math. USSR-Sb. 64 (1989), 129–140.

[We1] S. Webster, On the mapping problem for algebraic real hypersurfaces, Invent. Math. 43 (1977), 53–68.

[We2] S. Webster, Biholomorphic mappings and the Bergmann kernels off diagonal, Invent. Math. 51 (1979), 155–169.

[Z] A. Zygmund, Trignometric Series, Vol. I, Cambridge University Press, 2nd Edition, Cambridge 1968.

Department of Mathematics
Royal Institute of Technology
100 44 Stockholm, Sweden
ebenfelt@math.kth.se

Department of Mathematics
Rutgers University
New Brunswick, NJ 08903, U.S.A.
huangx@math.rutgers.edu

Compactness in the $\bar{\partial}$-Neumann problem*

Siqi Fu and Emil J. Straube

Contents

1. Introduction

Let Ω be a bounded pseudoconvex domain in $\mathbb{C}^n$. The $\bar{\partial}$ (or Dolbeault)-complex is the complex

$$L^2(\Omega) \xrightarrow{\bar{\partial}} L^2_{(0,1)}(\Omega) \xrightarrow{\bar{\partial}} L^2_{(0,2)}(\Omega) \xrightarrow{\bar{\partial}} \cdots \xrightarrow{\bar{\partial}} L^2_{(0,n)}(\Omega) \xrightarrow{\bar{\partial}} 0, \tag{1}$$

where $L^2_{(0,q)}(\Omega)$ denotes the space of $(0,q)$-forms on Ω with coefficients in $L^2(\Omega)$. The norm (with corresponding inner product) is $\| \sum_J' a_J \, d\bar{z}_J \|^2 = \sum_J' \int_\Omega |a_J|^2 dV$. Here, the prime denotes summation over strictly increasing q-tuples J, and $d\bar{z}_J = d\bar{z}_{j_1} \wedge d\bar{z}_{j_2} \wedge \cdots \wedge d\bar{z}_{j_q}$. The $\bar{\partial}$-operator acts via

$$\bar{\partial}\Big(\sum_J{}' a_J \, d\bar{z}_J\Big) = \sum_{j=1}^{n} \sum_J{}' \frac{\partial a_J}{\partial \bar{z}_j} d\bar{z}_j \wedge d\bar{z}_J . \tag{2}$$

*This work is supported in part by NSF grants DMS-9632028 and DMS-9801539.

The derivatives are taken in the distribution sense, and the domain of $\bar{\partial}$ consists of those $(0, q)$-forms where the right hand side is in $L^2_{(0,q+1)}(\Omega)$. $\bar{\partial}$ is then a densely defined closed operator, and so has an adjoint mapping $L^2_{(0,q+1)}(\Omega)$ into $L^2_{(0,q)}(\Omega)$. We denote this adjoint by $\bar{\partial}^*$. The complex Laplacian is the operator $\Box = \bar{\partial}\bar{\partial}^* + \bar{\partial}^*\bar{\partial}$, acting as an (unbounded) operator on $L^2_{(0,q)}(\Omega)$.

The basic L^2-existence theorem for the $\bar{\partial}$-Neumann problem (the problem of inverting $\Box$) goes back to Hörmander **[H65]**: for each q, $1 \le q \le n$, $\Box = \bar{\partial}\bar{\partial}^* + \bar{\partial}^*\bar{\partial}$ is an unbounded self-adjoint operator on $L^2_{(0,q)}(\Omega)$ which is surjective and (consequently) has a bounded inverse. This inverse is the $\bar{\partial}$-Neumann operator N_q. More precisely, N_q satisfies the estimate

$$\|N_q u\| \le \left(\frac{D^2 e}{q}\right) \|u\|, \quad u \in L^2_{(0,q)}(\Omega), \tag{3}$$

where D is the diameter of the domain Ω. If α is a $\bar{\partial}$-closed $(0, q + 1)$-form, then $\bar{\partial}^* N_{q+1}\alpha$ provides the solution orthogonal to the kernel of $\bar{\partial}$ (*i.e.*, the solution of minimal norm) to $\bar{\partial}u = \alpha$. The $\bar{\partial}$-Neumann operator is also closely related to the Bergman projection P_q, the orthogonal projection from $L^2_{(0,q)}$ onto its subspace consisting of $\bar{\partial}$-closed forms, via Kohn's formula $P_q = I - \bar{\partial}^* N_{q+1}\bar{\partial}$. We refer the reader to **[FK72, Kr92, BS99, CS01]** for further details and references on the L^2-theory of the $\bar{\partial}$-Neumann problem.

The question we consider in this article is that of compactness of N_q. This question is of interest for various reasons. We indicate some consequences of compactness in Section 2 below. In addition, from the point of view of finding necessary and sufficient conditions in terms of properties of the boundary, compactness appears to be more tractable than global regularity.

The compactness condition can be reformulated in several useful ways.

Lemma 1.1. *Let Ω be a bounded pseudoconvex domain, $1 \le q \le n$. Then the following are equivalent:*

(1) *The $\bar{\partial}$-Neumann operator N_q is compact from $L^2_{(0,q)}(\Omega)$ to itself.*

(2) *The embedding of the space* Dom $\bar{\partial} \cap$ Dom $\bar{\partial}^*$, *provided with the graph norm $u \to \|\bar{\partial}u\| + \|\bar{\partial}^*u\|$, into $L^2_{(0,q)}(\Omega)$ is compact.*

(3) *For every $\epsilon > 0$ there exists a constant $C_\epsilon > 0$ such that*

$$\|u\|^2 \le \epsilon\left(\|\bar{\partial}u\|^2 + \|\bar{\partial}^*u\|^2\right) + C_\epsilon \|u\|^2_{-1}$$

when $u \in$ Dom $\bar{\partial} \cap$ Dom $\bar{\partial}^*$.

(4) *The canonical solution operators $\bar{\partial}^* N_q : L^2_{(0,q)}(\Omega) \to L^2_{(0,q-1)}(\Omega)$ and $\bar{\partial}^* N_{q+1} : L^2_{(0,q+1)}(\Omega) \to L^2_{(0,q)}(\Omega)$ are compact.*

The statement in (3) is called a compactness estimate. In (4), we refer to $\bar{\partial}^* N_q$ and $\bar{\partial}^* N_{q+1}$ as "canonical solution operators", although strictly speaking the solution operators are the restrictions to the kernel of $\bar{\partial}$. However, $\bar{\partial}^* N_q$ and $\bar{\partial}^* N_{q+1}$ are zero on the orthogonal complement of ker $\bar{\partial}_q$ and ker $\bar{\partial}_{q+1}$, respectively. In particular, compactness is not affected. Note that saying that the canonical solution operator is compact is the same as saying that there exists *some* compact solution operator, as the projection onto the orthogonal complement of the kernel of $\bar{\partial}$ preserves compactness. The equivalence of (2) and (3) is in **[KN65]**, Lemma 1.1. That (1) is equivalent to (2) and (3) is an easy consequence of the general L^2-theory and the fact that $L^2_{(0,q)}(\Omega)$ embeds compactly into $W^{-1}_{(0,q)}(\Omega)$. Finally, the equivalence of (1) and (4) follows from the formula $N_q = (\bar{\partial}^* N_q)^*(\bar{\partial}^* N_q) + (\bar{\partial}^* N_{q+1})(\bar{\partial}^* N_{q+1})^*$ (see **[FK72]**, p. 55, **[Ra84]**).

It is useful to know that compactness is a *local* property. Roughly speaking, the $\bar{\partial}$-Neumann operator N_q on Ω is compact if and only if every boundary point has a neighborhood U such that the corresponding $\bar{\partial}$-Neumann operator on $U \cap \Omega$ is compact. For simplicity, we will assume that Ω and U are such that these intersections are domains (*i.e.*, connected); this is not a problem in the applications we have in mind (see Section 5).

Lemma 1.2. *Let Ω be a bounded pseudoconvex domain in $\mathbb{C}^n$, $1 \le q \le n$.*

(1) *If for every boundary point there exists a pseudoconvex domain U such that N_q on (the domain) $U \cap \Omega$ is compact, then N_q on Ω is compact.*

(2) *If U is smoothly bounded and strictly pseudoconvex and $U \cap \Omega$ is a domain, then if the $\bar{\partial}$-Neumann operator on $(0, q)$-forms on Ω is compact, so is the corresponding $\bar{\partial}$-Neumann operator on $U \cap \Omega$.*

The proof of (1) is straightforward and results from a partition of unity argument together with the interior elliptic regularity of $\bar{\partial} \oplus \bar{\partial}^*$. The proof of (2) is also standard, but it needs some ideas discussed in Section 3 below. Accordingly, we postpone this proof; it will be given at the end of Section 3.

2. Some consequences of compactness

In the case of smoothly bounded domains, a classical theorem of Kohn and Nirenberg **[KN65]** asserts that compactness of N_q (as an operator from $L^2_{(0,q)}(\Omega)$ to itself) implies global regularity in the sense of preservation of Sobolev spaces. We denote by $W^s(\Omega)$ the standard L^2-Sobolev spaces on Ω, and by $W^s_{(0,q)}(\Omega)$ the spaces of $(0, q)$-forms on Ω with coefficients in $W^s(\Omega)$.

Theorem 2.1. *Let Ω be a bounded pseudoconvex domain in $\mathbb{C}^n$ with smooth boundary. Let $1 \le q \le n$. If N_q is compact on $L^2_{(0,q)}(\Omega)$, then N_q is compact (in particular, continuous) as an operator from $W^s_{(0,q)}(\Omega)$ to itself, for all $s \ge 0$.*

Remark 1. In Theorem 2.1, the implication in the other direction is valid as well: if N_q is compact as an operator from $W^s_{(0,q)}(\Omega)$ to itself for *some* $s \ge 0$, then N_q is compact in $L^2_{(0,q)}(\Omega)$ (and hence in $W^s_{(0,q)}(\Omega)$, for all $s \ge 0$). This follows from an abstract theorem about compact operators over a Banach space which are symmetric with respect to a scalar product. Such operators are necessarily compact with respect to the norm given by the scalar product (see, e.g., **[La54]**, Corollary 2).

The Fredholm theory of Toeplitz operators is an immediate consequence of compactness of the $\bar\partial$-Neumann problem **[V72, HI97]**. In fact, compactness of the canonical solution operators implies that commutators between the Bergman projection and multiplication operators are compact **[CD97]**:

Proposition 2.2. *Let Ω be a bounded pseudoconvex domain in $\mathbb{C}^n$. Assume that for some $q, 0 \le q \le n-1$, the canonical solution operator $\bar\partial^* N_{q+1}$ is compact. Let M be a function that has bounded first order partial derivatives on Ω. Then the commutators $[P_q, M]$ between the Bergman projection P_q and the multiplication operator by M is compact on $L^2_{(0,q)}(\Omega)$.*

A short proof is as follows. For $f \in L^2_{(0,q)}(\Omega)$, let $f_1 := P_q f$, $f_2 := (I - P_q)f$. Since both projections are continuous, it suffices to see that $[P_q, M]$ is compact on both $\ker\bar\partial$ and $(\ker\bar\partial)^\perp$. We have

$$\begin{aligned}[P_q, M]f_1 &= P_q M f_1 - M P_q f_1 = M f_1 - \bar\partial^* N_{q+1}\bar\partial M f_1 - M f_1 \\ &= -\bar\partial^* N_{q+1}\bar\partial(M f_1) = -\bar\partial^* N_{q+1}(\bar\partial M \wedge f_1). \end{aligned} \tag{4}$$

(Compare **[Sa91]**, proof of Lemma 3 for a similar argument.) Since $\bar\partial M$ is bounded in Ω (so acts as an L^2-bounded multiplier) and since $\bar\partial^* N_{q+1}$ is compact, $[P_q, M]$ is compact on $\ker\bar\partial$. On the other hand,

$$\begin{aligned}[P_q, M]f_2 &= P_q M f_2 - M P_q f_2 = P_q M f_2 \\ &= P_q M \bar\partial^* N_{q+1}\bar\partial f_2 = P_q[M, \bar\partial^*](N_{q+1}\bar\partial) f_2. \end{aligned} \tag{5}$$

Here we have used the fact that $P_q \bar\partial^* M N_{q+1}\bar\partial f_2 = 0$, since the range of $\bar\partial^*$ is orthogonal to $\bar\partial$-closed forms (note that N_{q+1} maps into the domain of $\bar\partial^*$, which is preserved by multiplication with M). Since $[M, \bar\partial^*]$ acts as a zero order (bounded) operator on $N_{q+1}\bar\partial f_2$, and $N_{q+1}\bar\partial$ is compact (since $N_{q+1}\bar\partial = (\bar\partial^* N_{q+1})^*$), we are done.

Remark 2. In light of Lemma 1.1, (4), and Proposition 2.2, compactness of either N_q or N_{q+1} implies the compactness of $[P_q, M]$. It would be of interest to see to what extent properties of commutators between the Bergman projections and multiplication operators can be used to actually characterize compactness properties of the $\bar\partial$-Neumann problem. Compactness of the commutators $[P_q, M]$ does imply compactness of the canonical solution operator $\bar\partial^* N_{q+1}$ *restricted* to forms with *holomorphic* coefficients. This is essentially Lemma 2 in **[Sa91]**. Indeed, suppose that $u \in L^2_{(0,q+1)}(\Omega)$ has holomorphic coefficients. Write $u = \sum_j \sum'_K u_{jK} d\bar z_j \wedge d\bar z_K$, $1 \le j \le n$, $|K| = q$, then by (4) above, $-\sum_j [P_q, \bar z_j](\sum'_K u_{jK} d\bar z_K) = \sum_j \bar\partial^* N_{q+1}(\sum'_K u_{jK} d\bar z_j \wedge d\bar z_K) = \bar\partial^* N_{q+1} u$. Therefore, $\bar\partial^* N_{q+1}$ is compact on forms with holomorphic coefficients if the commutators $[P_q, \bar z_j]$, $1 \le j \le n$, are. On *convex* domains, compactness of this restriction *is* sufficient to give compactness of N_{q+1}, see **[FS98]**, Remark (2), Section 5.

Whether or not the $\bar\partial$-Neumann problem is compact has further ramifications for the theory of the Toeplitz C^*-algebras naturally associated to a domain, see **[Sa91, Sa95]**.

3. Sufficient conditions for compactness

We start with an inequality due to Catlin that is central to the subject (see **[Ca84a]**, p. 45; **[Ca87]**, Theorem 2.1; see also **[BS99]**, Section 2, for a somewhat different approach to this type of estimate). Assume for the moment that the boundary of Ω is sufficiently smooth, say C^2, to allow integration by parts. Let $\lambda \in C^2(\overline{\Omega})$ be normalized so that $0 \le \lambda \le 1$. Then

$$\sum_K{}' \sum_{j,k=1}^{n} \int_\Omega \frac{\partial^2 \lambda}{\partial z_j \partial \bar z_k} u_{jK} \bar u_{kK}\, dV \le e\big(\|\bar\partial u\|^2 + \|\bar\partial^* u\|^2\big) \tag{6}$$

for all $u \in \operatorname{Dom} \bar\partial \cap \operatorname{Dom} \bar\partial^* \subset L^2_{(0,q)}(\Omega)$. Heuristically, in view of (3) in Lemma 1.1 above (the compactness estimate), (6) shows that functions λ with "large" Hessians are desirable when compactness is the goal. To make this precise, it is convenient to recall the following notions (see **[Ca87]**).

Let $\Lambda_z^{(0,q)}$ be the space of $(0, q)$-forms at z equipped with the standard Hermitian metric $\big|\sum'_J u_J d\bar z_J\big|^2 = \sum'_J |u_J|^2$. For a C^2-smooth function $\lambda(z)$ in a neighborhood of z, let

$$H_q(\lambda)(z, u) = \sum_{|K|=q-1}{}' \sum_{j,k=1}^{n} \frac{\partial^2 \lambda(z)}{\partial z_j \partial \bar z_k} u_{jK} \bar u_{kK}, \quad u \in \Lambda_z^{(0,q)}.$$

Note that $H_1(\lambda)$ is the usual complex Hessian.

Lemma 3.1. *Let $M \ge 0$ be a constant. The following statements are equivalent.*

(1) $H_q(\lambda)(z, u) \ge M|u|^2$ *for all* $u \in \Lambda_z^{(0,q)}$.

(2) *For any orthonormal subset* $\{t^j,\ 1 \le j \le q\}$ *of* $\mathbb{C}^n$,

$$\sum_{j=1}^{q} H_1(\lambda)(z, t^j) \geq M.$$

(3) *The sum of any* q *eigenvalues of the Hermitian matrix* $(\partial^2\lambda/\partial z_j \partial \bar{z}_k)$ *is greater than or equal to* M.

The equivalence of (1) and (2) is in **[Ca87]**, pp. 189–190. The equivalence of (1) and (3) is implicit in **[H65]**, pp. 137. This lemma is proved by first establishing it in the case when the Hermitian matrix $(\partial^2\lambda/\partial z_j \partial \bar{z}_k)$ is diagonal, and then noting that statements (1)–(3) are invariant under a unitary transformation.

In light of Lemma 3.1, it is useful to have the following definition: For a bounded pseudoconvex domain Ω, we say that $b\Omega$ satisfies *property* (P_q) (or just (P_q)) if for every positive number M, there exists a neighborhood U of $b\Omega$ and a C^2-smooth function λ on $U \cap \Omega$, $0 \le \lambda \le 1$, such that for all z in $U \cap \Omega$, the sum of any q (equivalently: the smallest q) eigenvalues of the Hermitian form $\left(\partial\lambda/\partial z_j \partial \bar{z}_k\right)_{j,k=1}^{n}$ is at least M. This definition is from **[Ca84a]** (for $q = 1$). Note that $(P_1) \Rightarrow (P_2) \Rightarrow \cdots \Rightarrow (P_n)$. With this definition, we can make our heuristic statement from above precise.

Theorem 3.2. *Let* Ω *be a bounded pseudoconvex domain in* $\mathbb{C}^n$, $1 \le q \le n$. *If the boundary of* Ω *satisfies property* (P_q), *then the* $\bar{\partial}$*-Neumann operator* N_q *is compact.*

For sufficiently smooth domains, Theorem 3.2 is in **[Ca84a]**, proof of Theorem 1 (see also **[Ca87]**). The boundary regularity requirement was considerably weakened in **[HI97]**, Theorem 1. The second author showed in **[St97]**, Corollary 3, that no boundary regularity at all is needed.

Remark 3. The differentiability requirement on λ can be relaxed. For example, assume only that λ is continuous with its complex Hessian, interpreted as a *current*, bounded below by $Mdd^c|z|^2$. Then N_q is compact, $1 \le q \le n$. This is also contained in **[St97]**.

The proof of Theorem 3.2 results from the following considerations. Assume first that $b\Omega$ is smooth, so that we can apply (6). We note that by Lemma 3.1, the condition on λ (the function in the definition of property (P_q)) implies that

$$M|u(z)|^2 \le H_q(\lambda)(z, u) \tag{7}$$

for all $(0, q)$-forms u and $z \in U \cap \Omega$. Choosing $M \approx 1/\epsilon$, (7) combines with (6) to produce the desired compactness estimate, modulo terms that are compactly supported ((7) holds only near the boundary). These latter terms, however, are easily handled by the interior elliptic regularity of $\bar{\partial} \oplus \bar{\partial}^*$ (see **[Ca84a]**).

When no boundary regularity is assumed, one would like to obtain (6) by first integrating over approximating subdomains, and then passing to the limit. The reason that this is not immediate is that when forms are restricted to subdomains, they are in general no longer in the domain of $\bar{\partial}^*$ on the subdomain. This difficulty can be overcome by a regularization procedure introduced in **[St97]** that exploits the $\bar{\partial}$-Neumann operators of the subdomains. We refer the reader to **[St97]**, proof of Corollary 3, for details.

The simplest examples of domains satisfying (P_1) (hence (P_q), $1 < q$) are strictly pseudoconvex domains: it suffices to consider a strictly plurisubharmonic defining function. More generally, domains of finite type satisfy (P_1). This is far from obvious, however. It is a consequence of the analysis of finite type points in **[Ca84b]** and **[Da82]**, see **[Ca84a]**. But property (P_1) is considerably more general. It is not hard to see that if a domain is strictly pseudoconvex except at finitely many points, then it satisfies (P_1). In fact, if the infinite type points of the boundary have 2-dimensional Hausdorff measure zero, then the domain satisfies (P_1) (**[Bo88]**; **[Si87]**, Remark on p. 310). On the other hand, Sibony found examples of domains with (P_1) where the set of boundary points of infinite type has positive measure (**[Si87]**, p. 310).

Sibony undertook a systematic study of property (P_1) in the context of general compact sets in $\mathbb{C}^n$. One of the main tools is Choquet theory applied to the cone of plurisubharmonic functions. One can carry out an analogous study of property (P_q) by considering a cone of functions that reflects the condition on the Hessian used in Lemma 3.1, as in **[FS98]**. For an open set $U \subset \mathbb{C}^n$, denote by $P_q(U)$ the set of continuous functions λ on U such that for any $z \in U$ and orthonormal set of vectors $\{t_1, \ldots, t_q\}$ in $\mathbb{C}^n$, the function

$$\zeta = (\zeta_1, \ldots, \zeta_q) \in \mathbb{C}^q \mapsto \lambda(z + \zeta_1 t_1 + \cdots + \zeta_q t_q)$$

is subharmonic on $\{\zeta \in \mathbb{C}^q;\ z + \zeta_1 t_1 + \cdots + \zeta_q t_q \in U\}$. That is, $P_q(U)$ consists of the continuous functions on U that are subharmonic on each q-dimensional complex affine subspace. In particular, $P_1(U)$ is the set of all continuous plurisubharmonic functions and $P_n(U)$ is the set of all continuous subharmonic functions. $P_q(U)$ is a convex cone in $C(U)$ that is closed under taking the pointwise maximum of finitely many of its elements. Each function in $P_q(U)$ is a locally uniform limit of C^∞-smooth elements in P_q of slightly smaller open sets: this follows from the usual mollifier argument. It is easy to check that $-\sum_{j=1}^{q-1} |z_j|^2 + (q-1)\sum_{j=q}^{n} |z_j|^2 \in P_q(\mathbb{C}^n)$.

For a compact subset X of $\mathbb{C}^n$, let $P_q(X)$ be the uniform closure in $C(X)$ of functions that are in $P_q(U)$ for some neighborhood U of X. A probability measure μ on X is said to be a P_q-measure for $z \in X$ if

$$\lambda(z) \le \int_X \lambda\, d\mu, \qquad \text{for all } \lambda \in P_q(X).$$

Let $J_q(X)$ be the Choquet boundary of $P_q(X)$, *i.e.*, the subset of X consisting of points $z \in X$ such that the point mass is the only P_q-measure for z. We refer the reader to **[G78]**, Chapter 1 for the elements of Choquet theory. In particular, characterizations of the Choquet boundary are given in Theorem 1.13 in **[G78]**. We are interested in the situation where the Choquet boundary is all of X.

Proposition 3.3. *Let X be a compact subset of* $\mathbb{C}^n$. *Then the following are equivalent:*

(1) $J_q(X) = X$.

(2) $C(X) = P_q(X)$.

(3) *For any* $z \in X$, *there exists an* $r > 0$ *such that if* $X_r = X \cap \overline{B(z,r)}$ *then* $C(X_r) = P_q(X_r)$.

(4) *For any* $M > 0$, *there exists a function* λ *smooth in a neighborhood of* X *such that* $0 < \lambda(z) < 1$ *and* $H_q(\lambda)(z,u) \geq M|u|^2$ *for* $z \in X$ *and* $u \in \Lambda_z^{(0,q)}$.

Proposition 3.3 is from **[Si87]** (Proposition 1.3 and 1.4, compare also **[Si89]**) when $q = 1$, but the arguments cover the case of general q as well (in fact, many of the arguments hold in the general context of **[G78]**, Chapter 1). When $q = 1$, Sibony calls a compact set that has the (equivalent) properties in Proposition 3.3 B-regular. Accordingly, we shall use the term B_q*-regular* when $q \geq 1$. The significance of Proposition 3.3 in our context stems from (4): when $X = b\Omega$, where Ω is a bounded domain, then (4) is essentially property (P_q) for $b\Omega$. Actually, property (P_q) requires the function λ to exist only in a neighborhood of $b\Omega$ intersected with Ω, rather than in a full neighborhood of $b\Omega$. However, it is easy to see that on domains with relatively minimally regular boundary (for instance, when the boundary is locally a graph), the two notions coincide.

Remark 4. For boundaries of bounded domains which admit a bounded plurisubharmonic exhaustion function (hyperconvex domains), having the (equivalent) properties in Proposition 3.3 can also be characterized in terms of various properties related to the potential theory of the cone $P_q(\Omega)$, for example the existence of peak functions in $P_q(\Omega) \cap C(\overline{\Omega})$. For this, see **[Si87]**, Theorem 2.1, and **[Si89]**, Theorem 2.3, where the case $q = 1$ is treated. However, Sibony's arguments carry over to the case of general q.

The easiest way that B-regularity can fail is for the boundary to contain an analytic disc: in view of the maximum principle, $C(b\Omega) = P_q(b\Omega)$ cannot hold. A more direct argument is to pull back to the unit disc and to observe that subharmonic functions on the unit disc, with values between 0 and 1, cannot have arbitrarily large Hessians. Similarly, q-dimensional varieties in the boundary are incompatible with B_q-regularity. However, the absence of varieties is not sufficient for B_q-regularity in general: Sibony has given examples of complete smooth Hartogs domains in $\mathbb{C}^2$ whose boundaries contain no discs, yet are not B-regular. (**[Si87]**, p. 310). It turns out that on locally convexifiable domains, B_q-regularity (hence property (P_q)) *is* equivalent to the absence of q-dimensional varieties from the boundary; we will discuss this in Section 5 below.

B_q-regularity is preserved under countable unions.

Proposition 3.4. *Let X_k, $k = 1, 2, 3, \ldots$, be compact subsets of $\mathbb{C}^n$, all of which are B_q-regular. If $X = \bigcup_{k=1}^{\infty} X_k$ is compact, then X is B_q-regular.*

For $q = 1$, this is Proposition 1.9 in **[Si87]**. Sibony's proof works essentially verbatim to give the result for all q. (This includes the supporting Propositions 1.4 and 1.6, and Lemma 1.8.)

We conclude this section by completing the proof of Lemma 1.2 from Section 1. Assume that there is a compactness estimate for $(0, q)$-forms on Ω, and let U be strictly pseudoconvex (and $U \cap \Omega$ a domain). We show how to obtain a compactness estimate on $U \cap \Omega$. Let $u \in \operatorname{Dom} \bar{\partial} \cap \operatorname{Dom} \bar{\partial}^* \subset L^2_{(0,q)}(U \cap \Omega)$. Fix $\epsilon > 0$. The boundary of U is strictly pseudoconvex, so satisfies property (P_1), hence property (P_q) for all q. Thus for a smooth cutoff function ϕ_ϵ that is identically equal to 1 on bU and supported in a sufficiently small neighborhood of bU, we get by the arguments in the proof of Theorem 3.2:

$$\begin{aligned}
\|u\|^2_{U\cap\Omega} &\lesssim \|\phi_\epsilon u\|^2_{U\cap\Omega} + \|(1-\phi_\epsilon)u\|^2_{U\cap\Omega} \\
&\lesssim \epsilon\big(\|\bar{\partial}(\phi_\epsilon u)\|^2_{U\cap\Omega} + \|\bar{\partial}^*(\phi_\epsilon u)\|^2_{U\cap\Omega}\big) + \|(1-\phi_\epsilon)u\|^2_{U\cap\Omega} \\
&\lesssim \epsilon\big(\|\bar{\partial}u\|^2_{U\cap\Omega} + \|\bar{\partial}^*u\|^2_{U\cap\Omega}\big) + C_\epsilon\|\chi_\epsilon u\|^2_{U\cap\Omega} + \|(1-\phi_\epsilon)u\|^2_{U\cap\Omega}.
\end{aligned}$$

Here, χ_ϵ is a smooth cutoff function identically 1 on the support of $\nabla\phi_\epsilon$ whose support is compact in U, and $C_\epsilon := \max|\nabla\phi_\epsilon|$. Now view the forms $\chi_\epsilon u$ and $(1-\phi_\epsilon)u$ as forms on Ω; they are in the domain of $\bar{\partial}^*$ on Ω. Applying the compactness estimate on Ω, with an ϵ' *sufficiently small*, yields the desired compactness estimate on $U \cap \Omega$ (terms involving $\|u\|^2$ can be absorbed if ϵ' is chosen small enough). This completes the proof of Lemma 1.2.

4. Discs in the boundary vs. compactness

In this section, we restrict our attention to compactness of the $\bar{\partial}$-Neumann operator on $(0, 1)$-forms. Since a disc in the boundary is a blatant violation of B-regularity, and given that discs are known to prevent hypoellipticity (**[Ca81]**, **[DP81]**), it is natural to ask whether there is a corresponding failure of compactness.

Here is a simple example (taken from **[Li85]**, **[Kr88]**). Let Ω be the bidisc in $\mathbb{C}^2$, with the edge rounded. Denote by D the unit disc in $\mathbb{C}$, and let $f \in L^2(D) \cap \ker\bar{\partial}$. Using subscripts to denote L^2-norms on domains, we have that $\|\bar{z}_2 f(z_1)\|_\Omega \approx \|f\|_D \approx \|f\|_\Omega$. Also, $\bar{\partial}(\bar{z}_2 f(z_1)) = f(z_1)d\bar{z}_2$, and since $\bar{z}_2 f(z_1)$ is orthogonal in $L^2(\Omega)$ to the holomorphic functions, it is the canonical solution, *i.e.*, $\bar{z}_2 f(z_1) = \bar{\partial}^* N_1(f(z_1)d\bar{z}_2)$. Since the norms compare, compactness of $\bar{\partial}^* N_1$ (hence of N_1 by (4) in Lemma 1.1) would imply compactness of the unit ball in $L^2(D) \cap \ker\bar{\partial}$. Consequently, N_1 is not compact.

On the other hand, here is an example of a pseudoconvex Reinhardt domain in $\mathbb{C}^2$ with a disc in the boundary, but with compact $\bar{\partial}$-Neumann operator N_1. The domain is incomplete and non-smooth. Let $\Omega = \{(z_1, z_2) \in \mathbb{C}^2;\ |z_1|^2 + |z_2|^2 < 1,\ 0 < |z_1| < 1\}$, *i.e.*, Ω is the unit ball in $\mathbb{C}^2$ minus the variety $\{z_1 = 0\}$. The point is that the L^2-theory does not detect the deletion of this variety, and so is the same as on the unit ball. More precisely, the natural isometry $L^2_{(0,q)}(\Omega) \hookrightarrow L^2_{(0,q)}(B)$ (B denotes the unit ball) commutes with $\bar{\partial}_q$, $q = 0, 1, 2$. This can be checked by an argument completely analogous to that in **[Be82]**, p. 687. But then the $\bar{\partial}^*$ operators have similar commutation properties, hence so do the $\Box$ operators and their inverses, the $\bar{\partial}$-Neumann operators. In particular, N_1 on Ω inherits compactness from N_1 on B.

When some boundary regularity is assumed, the phenomenon in the above example cannot occur on domains in $\mathbb{C}^2$.

Proposition 4.1. *Let Ω be a bounded pseudoconvex domain in $\mathbb{C}^2$ with Lipschitz boundary. If the boundary of Ω contains an analytic disc, then the $\bar{\partial}$-Neumann operator N_1 on Ω is not compact.*

As of this writing, it is not known whether Proposition 4.1 holds in higher dimensions. Proposition 4.1 has been part of the folklore for many years; it is usually attributed to David Catlin; Michael Christ has also found a proof **[Ch98]**.

A proof of Proposition 4.1 results from adapting the ideas in **[Ca81]** and **[DP81]**. We have also used these ideas in **[FS98]**, Section 4. By Lemma 1.1, (4), it suffices to show that the canonical solution operator $\bar{\partial}^* N_1$ is not compact. We first assume that the disc in the boundary is an affine disc, say the disc $\overline{D_{2r_0}} \times \{0\}$, where $D_r := \{z \in \mathbb{C};\ |z| < r\}$. After a suitable complex linear change of coordinates, Ω is defined near the origin by $y_2 < \rho(z_1, x_2)$ ($z_j = x_j + iy_j$, $j = 1, 2$), where ρ is a Lipschitz function. Since $\rho(z_1, 0) = 0$ for $z_1 \in D_{2r_0}$, $|\rho(z_1, x_2)| = |\rho(z_1, x_2) - \rho(z_1, 0)| \le C|x_2|$ when $|x_2|$ is sufficiently small. Therefore, there exist circular wedges W_0 and W_1 of the same radius in the z_2-plane, symmetric about the y_2-axis, such that $D_{r_0} \times W_0 \subset \Omega$, and $\Omega \cap (D_{r_0} \times D_{r_1}) \subset D_{r_0} \times W_1$, for a suitable $r_1 > 0$ (shrink r_0 if necessary). We first observe that for any $r_3 > 0$ smaller than the (common) radius of W_0 and W_1, there is a sequence of holomorphic functions in W_1, bounded in $L^2(W_1)$, so that no subsequence converges in $L^2(W_0 \cap D_{r_3})$; *i.e.*, the restriction operator from the Bergman space on W_1 to the Bergman space on $W_0 \cap D_{r_3}$ is not compact. This is easily seen by considering the sequence

$$f_j(z_2) := \sqrt{\frac{2 - 2a_j}{\alpha}} R^{a_j - 1} z_2^{-a_j}$$

(defined via a branch cut along the positive imaginary axis), where $a_j \nearrow 1$. Here, R is the radius of the wedge W_1, and α is its angle. Denote by $\chi_1(t)$ a smooth cut-off function that is identically 1 for $0 \le t \le r_0/2$, and identically 0 for $t \ge 3r_0/4$, and by $\chi_2(t)$ a cut-off function that has these properties, but with r_1 instead of r_0. Then the forms $\alpha_j := \bar\partial(f_j(z_2)\chi_1(|z_1|)\chi_2(|z_2|))$ are $\bar\partial$-closed on Ω. Let $g_j := \bar\partial^* N_1 \alpha_j$. After passing to a subsequence, we may assume that $\{g_j\}_{j=1}^{\infty}$ converges in $L^2(\Omega)$ if $\bar\partial^* N_1$ is compact. Set $h_j(z_1, z_2) := \chi_1(|z_1|)\chi_2(|z_2|)f_j(z_2) - g_j(z_1, z_2)$. Then the h_j are holomorphic in Ω, and $h_j = -g_j$ outside the support of $\chi_1(|z_1|)\chi_2(|z_2|)$ (and hence $\{h_j\}_{j=1}^{\infty}$ converges in L^2 of Ω minus that support). Using the mean value property of holomorphic functions in the z_1 variable gives that $\{h_j\}_{j=1}^{\infty}$ also converges in $L^2(D_{r_0/2} \times (W_0 \cap D_{r_3}))$. This convergence also holds for $\{g_j\}_{j=1}^{\infty}$, hence for $\{\chi_1(|z_1|)\chi_2(|z_2|)f_j(z_2)\}_{j=1}^{\infty}$. But this implies that $\{f_j\}_{j=1}^{\infty}$ converges in $L^2(W_0 \cap D_{r_3})$ for any $r_3 < r_1/2$, a contradiction. Therefore, $\bar\partial^* N_1$ is not compact.

The general case of Proposition 4.1 is obtained by replacing the products disc $\times$ wedge above by suitable biholomorphic images of such products: the crucial non-compactness of the restriction operator between the Bergman spaces is invariant under these biholomorphisms.

Whether there can be obstructions to compactness (say for (0, 1)-forms) more subtle than discs in the boundary has been settled only recently. Peter Matheos **[M97]** showed that there exist smoothly bounded complete Hartogs domains in $\mathbb{C}^2$ without discs in their boundaries, but whose $\bar\partial$-Neumann operators are nonetheless not compact. Such a domain Ω will be of the form $\Omega = \{(z, w) \in \mathbb{C}^2;\ z \in \Omega_1, |w| < e^{-\Phi(z)}\}$, where Ω_1 is a domain in $\mathbb{C}$ and Φ is smooth and subharmonic in Ω_1. Assume that the boundary points of the form $(z, 0)$ are strictly pseudoconvex. The weakly pseudoconvex points are the points $\{(z, w) \in b\Omega;\ \Delta\Phi(z) = 0\}$. It is easy to see that $b\Omega$ will contain an analytic disc if and only if the set $W := \{z \in \Omega_1;\ \Delta\Phi(z) = 0\}$ has non-empty interior (to produce an analytic disc, consider a conjugate harmonic function to Φ on a (small) disc contained in the interior of W). The boundary will fail to satisfy property (P_1) (equivalently: B-regularity) if and only if W has non-empty *fine* interior, see **[Si87]**, p. 310. Recall that the fine topology is the smallest topology that makes all subharmonic functions continuous; see, *e.g.*, **[He69]** for properties of this topology. It is strictly larger than the Euclidean topology, and there exist compact sets with empty Euclidean interior, but non-empty fine interior. Matheos **[M97]** constructed examples of compact sets W with empty Euclidean interior and associated complete Hartogs domains whose weakly pseudoconvex boundary points project onto W and whose $\bar\partial$-Neumann operator N_1 is not compact. (Such sets W then necessarily have nonempty fine interior, by Sibony's result). Matheos' ideas combine with results from potential theory to give the following: *Every* compact set $W \subset \mathbb{C}$ with empty Euclidean interior but non-empty fine interior arises in this way.

Theorem 4.2. *Assume that W is a compact subset of $\mathbb{C}$ with non-empty fine interior. Then there exists a smooth, bounded, pseudoconvex, complete Hartogs domain Ω in $\mathbb{C}^2$ whose weakly pseudoconvex boundary points project onto W and whose $\bar\partial$-Neumann operator N_1 is not compact.*

Remark 5. In particular, when W has empty Euclidean interior, the complete Hartogs domain will have no discs in the boundary. Note that it does not follow from Theorem 4.2 that if the boundary of a smooth Hartogs domain does not satisfy property (P_1) (equivalently: the associated set W in the plane has non-empty fine interior), then its $\bar{\partial}$-Neumann operator N_1 is not compact. Whether or not this implication is true (that is, whether or not for complete smooth Hartogs domains in $\mathbb{C}^2$, property (P_1) and compactness of N_1 are actually equivalent) remains open.

Remark 6. Although the Hartogs domains in Theorem 4.2 have non-compact $\bar{\partial}$-Neumann operator, this operator is regular in Sobolev spaces; see **[BS92]**, Theorem 1: any bounded smooth Hartogs domain in $\mathbb{C}^2$ with no disc in the boundary is "nowhere wormlike", in the terminology of **[BS92]**, and so has globally regular $\bar{\partial}$-Neumann operator N_1.

Matheos first uses the machinery of Chang, Nagel, and Stein **[CNS92]** to show that a compactness estimate on Ω is equivalent to a compactness estimate on the boundary. This equivalence is established for all smooth pseudoconvex domains in $\mathbb{C}^2$. Then, the key property of the set W is that there exists a sequence of "Rayleigh functions" with supports shrinking to W. This last property turns out to hold for every compact set with non-empty fine interior (**[F99]**, Section 3). We now indicate how to combine this fact with Matheos' arguments to prove Theorem 4.2. We work on Ω directly; this results in some simplifications.

For an open set Ω in $\mathbb{R}^n$, let $\lambda(\Omega) = \inf\{\|\nabla f\|^2;\ \ f \in W_0^1(\Omega), \|f\| = 1\}$. The classical Rayleigh–Ritz formula says that $\lambda(\Omega)$ is the first eigenvalue of the Dirichlet problem for the (negative) Laplacian if Ω is regular with respect to the Dirichlet problem (see, e.g., **[Cha84]**). The Dirichlet problem has also been formulated for finely open sets in $\mathbb{R}^n$ (cf. **[F72]**). Among many properties, the monotonicity of λ with respect to the domain remains true for finely open sets. That is, if $\Omega_2 \supset \Omega_1 \neq \emptyset$ are finely open sets in $\mathbb{R}^n$, then $\lambda(\Omega_2) \leq \lambda(\Omega_1) < \infty$. We refer the reader to **[F99]**, Section 3, for a discussion of these (and further) results concerning the Dirichlet Laplacian on finely open sets.

Without loss of generality, we assume that W is contained in the open unit disc D. Let $\{D_j\}_{j=1}^{\infty}$ be the connected components of $D \setminus W$. Let $W_k = D \setminus \left(\bigcup_{j=1}^{k} \overline{D}_j\right)$. It follows from Theorem 10.14 in **[He69]** that $W_k \supset \operatorname{int}_f(W) \neq \emptyset$, where int_f denotes the interior in the fine topology. Therefore $\lambda(W_k) \leq \lambda(\operatorname{int}_f(W)) < \infty$. It follows that there exists a (Rayleigh) sequence of functions $v_k \in C_0^{\infty}(W_k)$ such that $\|v_k\| = 1$ and $\|\nabla v_k\| \lesssim 1$. (Throughout the rest of this section, $A(k) \lesssim B(k)$ means that $A(k) \leq C \cdot B(k)$, where C is some positive constant independent of k.)

Let φ be a non-negative smooth function on $\mathbb{C}$ that vanishes to infinite order on W and is strictly positive on $\mathbb{C} \setminus W$. Let $\psi_j = \varphi$ on $\overline{D}_j$ and $\psi_j = 0$ on $D \setminus \overline{D}_j$. Then $\varphi = \sum_{j=1}^{\infty} \psi_j$ on $\overline{D}$ and $\psi_j \in C^{\infty}(\overline{D})$. We now choose by induction a sequence $\{c_j\}_{j=1}^{\infty}$ in $(0,\ 1]$ and a strictly increasing sequence of positive integers $\{n_j\}_{j=1}^{\infty}$ such that the functions $\widetilde{\psi}_j = c_j \psi_j$ satisfy the following: (1) $n_j \int_{D_j} \widetilde{\psi}_j dA = 2\pi$; (2) $\|\widetilde{\psi}_{j+1}\|_{L^\infty} \leq \|\widetilde{\psi}_j\|_{L^\infty}$; (3) $n_j \|\widetilde{\psi}_{j+1}\|_{L^\infty} \leq 1$; (4) n_{j+1} is divisible by n_j. First choose n_1 sufficiently large so that $c_1 = 2\pi/(n_1 \int_{D_1} \psi_1 dA) \le 1$. Suppose that we have chosen c_j and n_j. We can choose an integer $m \ge 2$ sufficiently large so that $c_{j+1} = 2\pi/(m \cdot n_j \int_{D_{j+1}} \psi_{j+1} dA) \le \min\{1, \|\widetilde{\psi}_j\|_{L^\infty}/\|\psi_{j+1}\|_{L^\infty}, 1/(n_j\|\psi_{j+1}\|_{L^\infty})\}$. We then let $n_{j+1} = m \cdot n_j$.

We now let $\widetilde{\varphi} = \sum_{j=1}^\infty \widetilde{\psi}_j$. Then $\widetilde{\varphi} \in C^\infty(\overline{D})$. (The supports of $\widetilde{\psi}_j = c_j\psi_j$ are disjoint, except for points in W, where *all* derivatives vanish. Consequently the series actually converges in, for example, Sobolev norms.) Let $G = (1/2\pi)\log|z|$ and let $\Phi(z) = \int_D G(z-w)\widetilde{\varphi}(w)dA(w)$. Then $\Phi(z) \in C^\infty(\overline{D})$ and $\Delta\Phi = \widetilde{\varphi}$. Note that Φ is strictly subharmonic near ∂D. Extend Φ to the disc $\widetilde{D}$ with center 0 and radius 2 such that Φ is smooth on $\widetilde{D}$, strictly subharmonic on $\widetilde{D}\setminus D$, and equals $-(1/2)\log(4-|z|^2)$ near $\partial\widetilde{D}$. (Extend Φ smoothly to a function Φ_1 compactly supported in $\widetilde{D}$. Then add a radial function that is identically 0 for $|z| \le 1$ and agrees with $-(1/2)\log(4-|z|^2)$ when $|z|$ is close to 2. Such a function can be chosen to have its second (radial) derivative as big as we wish on a given compact subset of $\{1 < |z| < 2\}$, in particular on $\{\Delta\Phi_1 \le 0\}$.) Then $\Omega = \{(z,w);\ z \in \widetilde{D},\ |w| < \exp(-\Phi(z))\}$ is smooth. It remains to see that the $\bar\partial$-Neumann operator of Ω is not compact. For this, we shall construct a sequence of (0, 1)-forms that contradicts a compactness estimate, *i.e.*, we construct a sequence $\{u_k\}_{k=1}^\infty$ in $\mathrm{Dom}(\bar\partial^*) \cap C^\infty(\overline{\Omega})$ such that the estimate

$$\|u_k\|^2 \le \varepsilon(\|\bar\partial u_k\|^2 + \|\bar\partial^* u_k\|^2) + C_\varepsilon\|u_k\|_{-1}^2, \tag{7}$$

uniformly in k, will fail for some $\varepsilon > 0$.

Let $\widetilde{\varphi}_k = \sum_{j=1}^k \widetilde{\psi}_j$ and let $\Phi_k = \int_D G(z-w)\widetilde{\varphi}_k(w)dA(w)$. Then $\Delta\Phi_k = \widetilde{\varphi}_k = 0$ on W_k. It follows from the way we construct $\widetilde{\psi}_j$ that $n_k\Phi_k$ has a harmonic conjugate Θ_k on W_k whose values are determined up to an integer multiple of 2π. Therefore $\exp(i\Theta_k)$ is single valued and smooth on W_k. Let $f_k = \sqrt{n_k}v_k(z)\exp(n_k\Phi + i\Theta_k)w^{n_k-1}$ and let $u_k = f_k \cdot (\bar w d\bar z - 2\exp(-2\Phi)\Phi_z d\bar w)$. Then $u_k \in \mathrm{Dom}(\bar\partial^*) \cap C^\infty(\overline{\Omega})$ and $\|u_k\|^2 \approx \|f_k\|^2 \approx 1$. Moreover, $\{f_k\}$ and hence $\{u_k\}$ converge weakly to zero in $L^2(\Omega)$, and $L^2_{(0,1)}(\Omega)$, respectively (note that $f_j \perp f_k$ in $L^2(\Omega)$ when $j \ne k$) and so $\{u_k\}$ converges to zero in $W^{-1}_{(0,1)}(\Omega)$ in norm. (Alternatively, one easily checks that $\|u_k\|_{-1}^2 \approx \|f_k\|_{-1}^2 \lesssim (1/n_k^2)\|f_k\|^2 \approx 1/n_k^2$.) Therefore, in order to see that estimate (4) is violated for some $\epsilon > 0$, it suffices to see that $\|\bar\partial u_k\|^2 + \|\bar\partial^* u_k\|^2$ is bounded independently of k. The Kohn–Morrey formula gives

$$\|\bar\partial u_k\|^2 + \|\bar\partial^* u_k\|^2 = \int_{b\Omega} \sum_{j,\ell=1}^2 \frac{\partial^2\rho}{\partial z_j \partial\bar z_\ell}(u_k)_j\overline{(u_k)_\ell}\,d\sigma + \int_\Omega \sum_{j,\ell=1}^2 \left|\frac{\partial(u_k)_\ell}{\partial\bar z_j}\right|^2 dV,$$

where (temporarily) $z_1 = z$, $z_2 = w$. ρ is a defining function normalized so that $|\nabla\rho| = 1$ on $b\Omega$. We may take this function to be $w\bar w - \exp(-2\Phi(z))$, suitably normalized. Then the right hand side is dominated by a positive constant (independent of k) times

$$\int_{\partial\Omega} \Phi_{z\bar{z}}|f_k|^2\,d\sigma + \int_{\Omega}\left(|f_k|^2 + |f_{k\bar{z}}|^2\right)dV$$
$$\lesssim \int_{W_k}\left(n_k|v_k|^2\Phi_{z\bar{z}} + |v_k|^2 + |v_{k\bar{z}}|^2 + |v_k(n_k\Phi + i\Theta_k)_{\bar{z}}|^2\right)dA.$$

The integral of the first term in the last expression is not greater than $n_k\|\widetilde{\varphi}\|_{L^\infty(W_k)} \times \int_{W_k}|v_k|^2 dA = n_k\|\widetilde{\psi}_{k+1}\|_{L^\infty(W_k)} \leq 1$. The contribution from the second and third terms is under control because of our choice of the sequence $\{v_k\}_{k=1}^{\infty}$. The integral of the last term is not greater than $\|(n_k\Phi + i\Theta_k)_{\bar{z}}\|^2_{L^\infty(W_k)}$. Since $(n_k\Phi + i\Theta_k)_{\bar{z}} = n_k(\Phi - \Phi_k)_{\bar{z}}$ for $z \in W_k$ (because $(n_k\Phi_k + i\Theta_k)_{\bar{z}} = 0$), and the modulus of the latter term is not greater than

$$\frac{n_k}{4\pi}\int_D \frac{|\widetilde{\varphi} - \widetilde{\varphi}_k|}{|z-w|}\,dA(w) \lesssim n_k\|\widetilde{\varphi} - \widetilde{\varphi}_k\|_{L^\infty(D)} = n_k\|\widetilde{\psi}_{k+1}\|_{L^\infty(W_k)} \leq 1,$$

it follows that $\|\bar{\partial}u_k\|^2 + \|\bar{\partial}^* u_k\|^2 \lesssim 1$. This concludes the proof of Theorem 4.2.

5. Domains related to convex domains

The sufficient conditions for compactness discussed in Section 3 are also necessary in the case of convex domains, and, more generally, in the case of domains that are locally convexifiable. Moreover, for these domains, these conditions are equivalent to the absence of (germs of) varieties from the boundary. Pseudoconvex Reinhardt domains are "almost" in this latter class, that is,they are locally convexifiable at the boundary points away from the coordinate hyperplanes (but may not be so at boundary points on the coordinate hyperplanes, see, *e.g.*, **[FIK96]**), and some of the results carry over to this class.

We say that a domain is locally convexifiable if for every boundary point there is a neighborhood, and a biholomorphic map defined on this neighborhood, that takes the intersection of the domain with the neighborhood onto a convex domain. Note that boundaries of convex domains are locally graphs of Lipschitz functions, so the same is true for the boundary of a locally convexifiable domain. The following theorem comes from **[FS98]**.

Theorem 5.1. *Let Ω be a bounded pseudoconvex domain in $\mathbb{C}^n$ which is locally convexifiable, let $1 \leq q \leq n$. The following are equivalent:*

(1) *The $\bar{\partial}$-Neumann operator N_q is compact.*

(2) *The boundary of Ω does not contain any analytic variety of dimension $\geq q$.*

(3) *The boundary of Ω satisfies property (P_q).*

Note that for a locally convexifiable domain, property (P_q) is equivalent to B_q-regularity. Henkin and Iordan **[HI97]** had shown earlier that in the case of convex domains, if the boundary contains no analytic varieties of dimension ≥ 1, then all $\bar{\partial}$-Neumann operators N_q, $1 \leq q \leq n$, are compact.

Actually, that (2) $\Rightarrow$ (3) holds for the general case of locally convexifiable domains is only implicit in **[FS98]**. We now indicate how to make this explicit. By Proposition 3.3, property (P_q) for $b\Omega$ will follow if we show that for any $z \in b\Omega$, the point mass is the only P_q-measure on $b\Omega$ at z. This last property will follow if we can establish it locally, again by Proposition 3.3. So, let now K be a compact subset of the boundary, small enough so that it is contained in some open set on which there is a biholomorphism that convexifies the boundary, say $w = f(z) = (f_1(z), \ldots, f_n(z))$. By shrinking K if necessary, we may assume that there is a (small) domain U such that $K \subset U$, $f(U \cap \Omega)$ is convex, f is biholomorphic in a neighborhood of $\overline{U \cap \Omega}$, and there is no variety of dimension greater than or equal to q in the boundary of $f(U \cap \Omega)$. By **[FS98]**, Proposition 3.2, there exists, through each boundary point, a complex affine subspace L of dimension $\leq q - 1$ such that $L \cap \overline{f(U \cap \Omega)}$ is a peak set for $A(f(U \cap \Omega))$, the algebra of functions analytic in $f(U \cap \Omega)$ and continuous on $\overline{f(U \cap \Omega)}$. The inverse image under f of L gives a manifold of dimension $\leq q - 1$ such that $f^{-1}(L) \cap \overline{U \cap \Omega}$ is a peak set for $A(U \cap \Omega)$. For every boundary point in K, there exists a peak set for $A(U \cap \Omega)$ of this form that contains the point. Let g be the corresponding peak function. Note that every element in $A(U \cap \Omega)$ can be approximated uniformly on $\overline{U \cap \Omega}$ by functions analytic in a neighborhood of $\overline{U \cap \Omega}$, since this is true for the corresponding algebra on the *convex* domain $f(U \cap \Omega)$, and f is biholomorphic in a neighborhood of $\overline{U \cap \Omega}$. Thus $|g| \in P_q(K)$. Consequently, for $z \in K$, all P_q-measures of K at z are supported on $K \cap f^{-1}(L)$. Let now $z_0 \in K$, and choose coordinates in $\mathbb{C}^n$ such that $f(z_0) = 0$ and $L \subset \{w_q = w_{q+1} = \cdots = 0\}$. Then for sufficiently large M, the function $-\sum_{j=1}^{q-1} |f_j(z)|^2 + M \sum_{j=q}^{n} |f_j(z)|^2$ is in $P_q(K)$. It equals zero at z_0, and is non-positive on $K \cap f^{-1}(L)$. If we integrate against a P_q-measure for z_0 over K, and take into account that the measure is supported on $K \cap f^{-1}(L)$, it follows that it is actually a point mass at z_0.

Remark 7. The proof of (1) $\Rightarrow$ (2) in **[FS98]** establishes the following: if N_q is compact and Ω is convexifiable at (hence near) a boundary point p, then the boundary cannot contain q-dimensional varieties near p. The reduction to the (globally) convex case uses the localization in Lemma 1.2 in an essential way, together with the observation that compactness of the relevant $\bar{\partial}$-Neumann operators is invariant under the convexifying biholomorphism (Lemma 1.1, (4), and the observation that compactness of the canonical solution operator follows once there is some compact solution operator), see **[FS98]**, Remark (1) in Section 5.

Remark 8. On *convex* domains, more is true. As mentioned in Section 2, compactness of the commutators $[P_{q-1}, \bar{z}_j]$ implies compactness of $\bar{\partial}^* N_q$ restricted to forms with holomorphic coefficients (always), and this in turn implies compactness of $\bar{\partial}^* N_q$ on all of $L^2_{(0,q)}(\Omega)$ if Ω is convex (**[FS98]**, Remark (2), Section 5). Thus in the convex case, properties (1)–(3) in Theorem 11 are furthermore equivalent to compactness of the commutators $[P_{q-1}, M]$ discussed in Proposition 2.2 as well as to compactness of $\bar\partial^* N_q$, restricted to forms with holomorphic coefficients. In addition, the absence of varieties from the boundary is equivalent with the absence of affine varieties. This is a simple manifestation of the general principle that on convex boundaries, questions of orders of contact of varieties are decided by orders of contact of affine varieties. (See **[FS98]**, Section 2, and the references given there.)

As mentioned earlier, pseudoconvex Reinhardt domains are locally convexifiable at most of their boundary points, so it is not surprising that the above ideas also give results on this class of domains.

Theorem 5.2. *Let Ω be a bounded pseudoconvex Reinhardt domain in $\mathbb{C}^n$, $1 \le q \le n$. If the boundary of Ω does not contain an analytic variety of dimension $\ge q$, then it is B_q-regular, and consequently, the $\bar\partial$-Neumann operator N_q on $(0, q)$-forms is compact.*

The fact that if the boundary of a Reinhardt domain contains no varieties of dimension ≥ 1, then N_q, $1 \le q \le n$, is compact is in **[HI97]**, Corollary 1. However, there is an inaccuracy in the proof: it is asserted that on a bounded pseudoconvex Reinhardt domain Ω, every boundary point is a peak point for the algebra $A(\Omega)$, the algebra of continuous functions on $\overline{\Omega}$ which are holomorphic on Ω. Pseudoconvex Reinhardt domains in $\mathbb{C}^2$ lying over the punctured disc show that this is not the case in general: the boundary point corresponding to the puncture is not a peak point. Likewise, such a domain will not satisfy property (B) from **[HI97]**.

Remark 9. The example of the domain $\Omega = \{(z_1, z_2) \in \mathbb{C}^2;\ |z_1|^2 + |z_2|^2 < 1,\ 0 < |z_1| < 1\}$ discussed in Section 4 above shows that compactness of the $\bar\partial$-Neumann operator N_q on a Reinhardt domain does in general *not* imply the absence of varieties from the boundary. It does imply, in view of Remark 7, that *the boundary intersected with the complement of the coordinate hyperplanes cannot contain q-dimensional varieties if N_q is compact.* (The implication that if N_1 is compact, then the boundary (away from the coordinate hyperplanes) cannot contain analytic discs is contained in work of Salinas concerning Toeplitz C^*-algebras on Bergman spaces. For this work, and further references, see the two surveys **[Sa91, Sa95]**.)

We now prove Theorem 5.2. The proof is by induction on the dimension n, as follows. When $n = 1$, Ω is either a disk, a punctured disk, or an annulus. In all three cases, the boundary is B-regular. We assume now that the theorem holds for all dimensions $\le n - 1$, and show that it then holds in dimension n. Since the domain Ω is pseudoconvex, the boundary is locally convexifiable near each boundary point $(z_1, \ldots, z_n)$ where none of the z_j $(1 \le j \le n)$ are 0. As in the proof of Theorem 5.1 (the implication (2) $\Rightarrow$ (3)), it follows that for every such boundary point z, there is $r > 0$ such that the compact set $b\Omega \cap \overline{B(z, r)}$ is B_q-regular. Consequently, the part of the boundary where all coordinates are non-zero (this is not a compact set) is a countable union of compact sets, all of which are B_q-regular.

The rest of the boundary is a finite union of compact sets of the form $b\Omega \cap \{z_{j_1} = \cdots = z_{j_k} = 0, z_{j_{k+1}} \neq 0, \dots, z_{j_n} \neq 0\}$, where $\{j_1, j_2, \dots, j_n\}$ is a permutation of $\{1, 2, \dots, n\}$ and k ranges from 1 to $n-1$. (The case $k = n$ results in only one set, $b\Omega \cap \{0\}$, which is either empty or $\{0\}$. But the set $\{0\}$ is B_q-regular.) So fix $(j_1, \dots, j_k)$. Assume first that $n - k \geq q$. We will show that $b\Omega \cap \{z_{j_1} = \cdots = z_{j_k} = 0, z_{j_{k+1}} \neq 0, \dots, z_{j_n} \neq 0\}$ (which we may assume nonempty) is the boundary, in $\{z_{j_1} = \cdots = z_{j_k} = 0\} \equiv \mathbb{C}^{n-k}$, of the bounded Reinhardt domain $\Omega \cap \{z_{j_1} = \cdots = z_{j_k} = 0\}$. Since Ω is pseudoconvex, its logarithmic image, $\log|\Omega| := \{(\log|z_1|, \dots, \log|z_n|);\ z \in \Omega, z_j \neq 0, 1 \leq j \leq n\}$, is convex. Since there is a boundary point with a zero coordinate, it is unbounded. We use the tools and notions developed in Section 8 of **[Ro70]**. Let $\{z_\nu\}_{\nu=1}^{\infty}$ be a sequence of points in Ω that converges to a boundary point $(0, \dots, 0, z_{k+1}, \dots, z_n)$, with $z_{k+1} \neq 0, \dots, z_n \neq 0$. (Without loss of generality, we assume that $(j_1, j_2, \dots, j_n) = (1, 2, \dots, n)$.) It follows from the convexity of the unbounded set $\log|\Omega|$ (specifically, see **[Ro70]**, Theorem 8.2 and Theorem 8.3) that there is a vector $r = (r_1, \dots, r_k, 0, \dots, 0) \in \mathbb{R}^n$ such that $\zeta + r$ also belongs to $\log|\Omega|$ for each $\zeta \in \log|\Omega|$ (r is in the recession cone of $\log|\Omega|$ in the terminology of **[Ro70]**). This shows that the projection of Ω onto $\{z_1 = \cdots = z_k = 0\}$ consists entirely of points in $\overline{\Omega}$. Since the boundary of Ω contains no q-dimensional varieties, and since $n-k \geq q$, at least one of these points has to be an interior point of Ω. Pseudoconvexity now implies that Ω is complete with respect to the variables $(z_1, \dots, z_k)$. Therefore, the projection of Ω onto $\{z_1 = \cdots = z_k = 0\}$ equals the intersection of Ω with this coordinate "plane". Consequently, $b\Omega \cap \{z_1 = \cdots = z_k = 0\}$ is the same as the boundary of $\Omega \cap \{z_1 = \cdots = z_k = 0\}$. This is the boundary of a pseudoconvex bounded Reinhardt domain in $\mathbb{C}^{n-k}$. It contains no q-dimensional varieties. By the induction assumption, it is B_q-regular in $\mathbb{C}^{n-k}$. By considering functions of the form $\lambda(z_{k+1}, \dots, z_n) + M(|z_1|^2 + \cdots + |z_k|^2)$, one sees that $b\Omega \cap \{z_1 = \cdots = z_k = 0\}$ also is B_q-regular as a subset of $\mathbb{C}^n$.

It remains to consider the case $n - k < q$. In this case, any compact subset of $\{z_1 = \cdots = z_k = 0\}$ is B_q-regular as a subset of $\mathbb{C}^n$ (consider functions in $P_q(\mathbb{C}^n)$ of the form $M(|z_1|^2 + \cdots + |z_k|^2)$).

We now know that $b\Omega$ is a countable union of compact, B_q-regular subsets of $\mathbb{C}^n$. Therefore, by Proposition 3.4, $b\Omega$ itself is B_q-regular as well.

Remark 10. Recall the second example from Section 4 above: Ω is the unit ball in $\mathbb{C}^2$ minus the variety $\{z_1 = 0\}$, and N_1 is compact on Ω. This case is not covered by Theorem 5.2 (the boundary contains a disc). It is instructive to compare the two situations. In the above proof of Theorem 5.2, we have used the fact that $b\Omega$ contains no q-dimensional variety twice. Namely, we used it first to conclude that the projection of Ω onto $\{z_1 = \cdots = z_k = 0\}$ consists of interior points of Ω. This use of the absence of varieties is merely convenient: one could "fill in" these points as in the example. However, in order to make the induction run, we need to know that the *boundary* of this projection is free of q-dimensional varieties, and this *is* crucial for the argument, unless $q = n - 1$. When $q = n - 1$, the latter property comes for free. From these ideas, there results the following characterization of compactness of N_{n-1} on a bounded pseudoconvex Reinhardt domain Ω in $\mathbb{C}^n$: N_{n-1} *is compact if and only if* $b\Omega \setminus \{z_1 \dots z_n = 0\}$ *contains no* $(n-1)$*-dimensional variety*. For complete Reinhardt domains, this characterization was obtained in **[SSU89]**.

Acknowledgments. Part of this work was done while the first author visited Texas A & M University in the summer of 1999. He thanks the Department of Mathematics and Harold and Heidi Boas for their hospitality.

References

[Be82] Steven R. Bell, The Bergman kernel function and proper holomorphic mappings, Trans. Amer. Math. Soc. 270 (1982), 585–591.

[Bo88] Harold P. Boas, Small sets of infinite type are benign for the $\bar{\partial}$-Neumann problem, Proc. Amer. Math. Soc. 103 (1988), 569–578.

[BS92] Harold P. Boas and Emil J. Straube, The Bergman projection on Hartogs domains in $\mathbb{C}^2$, Trans. Amer. Math. Soc. 331 (1992), 529–540.

[BS99] —, Global regularity of the $\bar{\partial}$-Neumann problem: a survey of the L^2-Sobolev theory, in: Several Complex Variables (M. Schneider and Y.-T. Siu, eds.), Math. Sci. Res. Inst. Publ. 37, Cambridge University Press, 1999.

[Ca81] David Catlin, Necessary conditions for subellipticity and hypoellipticity for the $\bar{\partial}$-Neumann problem on pseudoconvex domains, in: Recent Developments in Several Complex Variables (John E. Fornæss, ed.), Ann. of Math. Stud. 100, Princeton University Press, 1981, 93–100.

[Ca83] —, Necessary conditions for subellipticity of the $\bar{\partial}$-Neumann problem, Ann. of Math. (2) 117 (1983), 147–171.

[Ca84a] —, Global regularity of the $\bar{\partial}$-Neumann problem, in: Complex Analysis of Several Variables (Yum-Tong Siu, ed.), Proc. Symp. Pure Math. 41, Amer. Math. Soc., Providence, RI, 1984, 39–49.

[Ca84b] —, Boundary invariants of pseudoconvex domains, Ann. of Math. (2) 120 (1984), 529–586.

[Ca87] —, Subelliptic estimates for the $\bar{\partial}$-Neumann problem on pseudoconvex domains, Ann. of Math. (2) 126 (1987), 131–191.

[CD97] David Catlin and John D'Angelo, Positivity conditions for bihomogeneous polynomials, Math. Res. Lett. 4 (1997), 555–567.

[CNS92] D.-C. Chang, A. Nagel, and E. M. Stein, Estimates for the $\bar{\partial}$-Neumann problem in pseudoconvex domains of finite type in $\mathbb{C}^2$, Acta Math. 169 (1992), 153–228.

[Cha84] Isaac Chavel, Eigenvalues in Riemannian geometry, Pure Appl. Math. 115, Academic Press, 1984

[CS01] So-Chin Chen and Mei-Chi Shaw, Partial differential equations in several complex variables, Amer. Math. Soc./International Press, 2001.

[Ch98] Michael Christ, Private communication, 1998.

[Da82] John D'Angelo, Real hypersurfaces, orders of contact, and applications, Ann. of Math. (2) 115 (1982), 615–637.

[DP81] Klas Diederich and Peter Pflug, Necessary conditions for hypoellipticity of the $\bar{\partial}$-problem, in: Recent Developments in Several Complex Variables (John E. Fornæss, ed.), Ann. of Math. Stud. 100, Princeton University Press, 1981, 151–154.

[FK72] G. B. Folland and J. J. Kohn, The Neumann problem for the Cauchy–Riemann complex, Ann. of Math. Stud. 75, Princeton University Press, 1972.

[FIK96] Siqi Fu, Alexander Isaev, Steven Krantz, Finite type conditions on Reinhardt domains, Complex Variables Theory Appl. 31 (1996), 357–363.

[FS98] Siqi Fu and Emil J. Straube, Compactness of the $\bar{\partial}$-Neumann problem on convex domains, J. Funct. Anal. 159 (1998), 629–641.

[F72] Bent Fuglede, Finely Harmonic Functions, Lecture Notes in Math. 289, Springer-Verlag, 1972.

[F99] —, The Dirichlet Laplacian on finely open sets, Potential Anal. 10 (1999), 91–101.

[G78] Theodore W. Gamelin, Uniform algebras and Jensen measures, London Math. Soc. Lecture Note Ser. 32, Cambridge University Press, 1978.

[He69] L. L. Helms, Introduction to potential theory, Wiley-Interscience, 1969.

[HI97] G. M. Henkin and A. Iordan, Compactness of the Neumann operator for hyperconvex domains with non-smooth B regular boundary, Math. Ann. 307 (1997), 151–168.

[H65] Lars Hörmander, L^2 estimates and existence theorems for the $\bar{\partial}$ operator, Acta Math. 113 (1965), 89–152.

[KN65] J. J. Kohn and L. Nirenberg, Non-coercive boundary value problems, Comm. Pure Appl. Math. 18 (1965), 443–492.

[Kr88] Steven G. Krantz, Compactness of the $\bar{\partial}$-Neumann operator, Proc. Amer. Math. Soc. 103 (1988), 1136–1138.

[Kr92] —, Partial differential equations and complex analysis, CRC Press, Boca Raton, FL, 1992.

[La54] Peter D. Lax, Symmetrizable linear transformations, Comm. Pure Appl. Math. 7 (1954), 633–647.

[Li85] E. Ligocka, The regularity of the weighted Bergman projections, in: Seminar on deformations (J. Lawrynowicz, ed.), Lecture Notes in Math. 1165, Springer-Verlag, 1985, 197–203.

[M97] Peter Matheos, A Hartogs domain with no analytic discs in the boundary for which the $\bar{\partial}$-Neumann problem is not compact, preprint, 1997, to appear in J. Geom. Anal.

[Ra84] R. Michael Range, The $\bar{\partial}$-Neumann operator on the unit ball in $\mathbb{C}^n$, Math. Ann. 266 (1984), 449–456.

[Ro70] R. Tyrrell Rockafellar, Convex analysis, Princeton Math. Ser. 28, Princeton University Press, Princeton, NJ, 1970.

[Sa91] Norberto Salinas, Noncompactness of the $\bar{\partial}$-Neumann problem and Toeplitz C^*-algebras, in: Several Complex Variables and Complex Geometry (Eric Bedford, John P. D'Angelo, Robert E. Greene, and Steven G. Krantz, eds.), vol. 3, Proc. Sympos. Pure Math. 52, Amer. Math. Soc., Providence, RI, 1991, 329–334.

[Sa95] —, Toeplitz C^*-algebras and several complex variables, in Multivariable operator theory, Contemp. Math. 185, Amer. Math. Soc., Providence, RI, 1995.

[SSU89] N. Salinas, A. Sheu, and H. Upmeier, Toeplitz operators on pseudoconvex domains and foliation C^*-algebras, Ann. of Math. 130 (1989), 531–565.

[Si87] N. Sibony, Une classe de domaines pseudoconvexes, Duke Math. J. 55 (1987), 299–319.

[Si89] —, Some aspects of weakly pseudoconvex domains, in: Several Complex Variables and Complex Geometry (Eric Bedford, John P. D'Angelo, Robert E. Greene, and Steven G. Krantz, eds.), vol. 1, Proc. Sympos. Pure Math. 52, Amer. Math. Soc., Providence, RI, 1991, 119–232.

[St97] Emil J. Straube, Plurisubharmonic functions and subellipticity of the $\bar{\partial}$-Neumann problem on non-smooth domains, Math. Res. Lett. 4 (1997), 459–467.

[V72] U. Venugopalkrishna, Fredholm operators associated with strongly pseudoconvex domains in $\mathbb{C}^n$, J. Funct. Anal. 9 (1972), 349–373.

University of Wyoming
Department of Mathematics, Laramie, WY 82071, U.S.A.
sfu@uwyo.edu

Texas A & M University
Department of Mathematics, College Station, TX 77843, U.S.A.
straube@math.tamu.edu

Hypoellipticity at non-subelliptic points

*Joseph J. Kohn**

Abstract. Here we deal with pseudoconvex domains and CR manifolds which are subelliptic outside of a set S. We wish to find geometric conditions on S such that hypoellipticity holds on points of S for the $\bar{\partial}$-Neumann problem and for $\Box_b$, respectively. In particular we show that if S consists of a finite number of points then hypoellipticity holds.

1. Introduction

Let M be a pseudoconvex CR manifold of dimension $2n-1$. If $x_o \in M$ we say that $\Box_b$ is *hypoelliptic* at x_o if there exists a neighborhood U of x_o such that if φ and α are forms on M with L_2 coefficients and $\alpha|_U$ is C^∞ with $\Box_b\varphi = \alpha$ then $\varphi|_U \in C^\infty$. Hypoellipticity at x_o is a consequence of subellipticity at x_o (see **[KN]**). The operator $\Box_b$ is *subelliptic* at x_o if there exist a neighborhood U and positive constants ε and C such that

$$\|\varphi\|_\varepsilon^2 \le C(\Box_b\varphi, \varphi) = CQ_b(\varphi, \varphi), \tag{1.1}$$

for all $\varphi \in C_0^\infty(U)$.

Subelliptic estimates for the $\bar{\partial}$-Neumann problem at a point $x_o \in b\Omega$, where $\Omega \subset \mathbb{C}^n$ is a pseudoconvex domain with smooth boundary $b\Omega$, are defined as follows. Suppose that there exist a neighborhood U and positive constants ε and C such that

$$\|\varphi\|_\varepsilon^2 \le C(\|\bar{\partial}\varphi\|^2 + \|\bar{\partial}^*\varphi\|^2) = CQ(\varphi, \varphi), \tag{1.2}$$

for all $\varphi \in C_0^\infty(U \cap \bar{\Omega})$ with φ in the domain of $\bar{\partial}^*$. The necessary and sufficient condition for such an estimate to hold for the $\bar{\partial}$-Neumann problem near a point in the boundary of a pseudoconvex domain is that the D'Angelo type at the point is finite (see **[D]** and **[C]**). The estimates (1.1) and (1.2) are closely related (see **[K6]**), in fact, if $M = b\Omega$ then (1.2) is equivalent to

$$\|\varphi^+\|_\varepsilon^2 \le CQ_b(\varphi, \varphi), \tag{1.3}$$

for all $\varphi \in C_0^\infty(U)$, where $\varphi = \varphi^+ + \varphi^0 + \varphi^-$ denotes a certain microlocal decomposition which will be defined below.

*Research was partially supported by NSF Grant DMS#9801626

The problem is to find conditions under which hypoellipticity holds when subellipticity fails. The most desirable condition is a purely geometric condition such as in Fediĭ's theorem (see **[F]**), which states that the operator $\frac{\partial^2}{\partial x^2} + a(x)\frac{\partial^2}{\partial y^2}$ with $a \in C^\infty$ is hypoelliptic if $a(x) > 0$ when $x \neq 0$. In **[K6]** Fediĭ's result is generalized to degenerate subelliptic operators in any dimension. This phenomenon appears on CR manifolds, thus on a pseudoconvex CR manifold $\Box_b$ is hypoelliptic in a neighborhood U of x_o if it is subelliptic at all x with $x \neq x_o$. The natural geometric condition for hypoellipticity seems to be that the points where subellipticity fails are contained in a real curve transversal to the "good" directions (i.e., the tangent vectors to the curve do not lie in $T^{(1,0)} + T^{(0,1)}$), this is true for certain classes of CR manifolds (see **[K5]** and **[K7]**) but Christ has shown that it is false in general (see **[Ch2]**, **[Ch3]**). The example of Kusuoka and Stroock (see **[KS]**) illustrates that, in general, the geometry alone of the set on which the operator degenerates does not give sufficient information to determine whether it is hypoelliptic. They study the operator $\frac{\partial^2}{\partial x^2} + a^2(x)\frac{\partial^2}{\partial y^2} + \frac{\partial^2}{\partial z^2}$ in $\mathbb{R}^3$ with $a(x) \neq 0$ if $x \neq 0$ and prove that it is hypoelliptic if and only if $\lim_{x\to 0} x \log a(x) = 0$. Hypoellipticity of such operators is obtained by use of superlogarithmic estimates (see **[Ch1]**, **[M]**, and **[K6]**). Here we will assume that subelliptic estimates hold at all points outside a subset S of M and find conditions on S which insure that $\Box_b$ is hypoelliptic. The main result is that $\Box_b$ is hypoelliptic if S consists of isolated points.

One of the crucial steps in proving these results is the following lemma, proved in **[K6]**.

Lemma 1.1. *Let M be a pseudoconvex CR manifold of dimension $2n-1$, let $x_o \in M$ and let λ be a real non-negative C^∞ function such that $\lambda(x_o) = 0$ and $\mathrm{Re}\big(L_i \bar{L}_j(\lambda)|_{x_o}\big)$ is positive-definite. Let $\sigma_s = e^{s\lambda}$ Then there exists a neighborhood U of x_o and $C > 0$ such that*

$$\|\sigma_s\zeta\varphi^+\|^2 \le C\Big(\frac{1}{s}\max_{\mathrm{supp}\,\zeta} \sigma_{2s}(Q_b(\zeta\varphi^+,\zeta\varphi^+) + \|\zeta\varphi^+\|^2) + s\max_{\mathrm{supp}\,\zeta} |L_i(\lambda)|^2\sigma_{2s}\|\zeta\varphi^+\|^2 + \|\varphi^+\|^2_{-1}\Big) \quad (1.4)$$

where $\zeta \in C_0^\infty(U)$ and $\varphi \in \mathcal{A}_b^{0,1}$ with $\mathrm{supp}(\varphi) \subset U'$, *where $U' \supset \bar{U}$.*

2. Definitions and preliminaries

Definition 2.1. Let M be a $2n-1$ dimensional differentiable manifold. Denote by $\mathbb{C}T(M)$ the complexified tangent bundle of M. An *integrable CR structure* on M is given by a subbundle $T^{1,0} \subset \mathbb{C}T(M)$ with the following properties. If $T^{0,1}$ denotes the conjugate of $T^{1,0}$ then $T^{1,0} \cap T^{0,1} = \{0\}$. The subbundle $T^{1,0} \oplus T^{0,1}$ is of codimension one. If L and L' are local vectorfields with values in $T^{1,0}$ then the commutator $[L, L'] = LL' - L'L$ also has values in $T^{1,0}$. We say M is a *CR manifold* if M has a given integrable CR structure.

Definition 2.2. A *form of type* (p, q) at $x \in M$ is a skew-symmetric multilinear map

$$\varphi : \underbrace{T_x^{1,0} \times \cdots \times T_x^{1,0}}_{p} \times \underbrace{T_x^{0,1} \times \cdots \times T_x^{0,1}}_{q} \to \mathbb{C}.$$

The *bundle of* (p, q)*-forms* is denoted by $\mathcal{A}^{p,q}$.

Definition 2.3. The operator $\bar{\partial}_b : \mathcal{A}_b^{p,q} \to \mathcal{A}_b^{p,q+1}$ is defined as follows. If $\varphi \in \mathcal{A}^{p,q}$ let φ' be a (p, q)-form which restricted to $\prod_p T^{1,0} \times \prod_q T^{0,1}$ equals φ. Then $\bar{\partial}_b\varphi$ is the restriction of $d\varphi'$ to $\prod_p T^{1,0} \times \prod_{q+1} T^{0,1}$. The operator $\bar{\partial}_b^* : \mathcal{A}_b^{p,q} \to \mathcal{A}_b^{p,q-1}$ is the L_2-adjoint of $\bar{\partial}_b$.

Definition 2.4. Denote by θ a non-vanishing real 1-form which annihilates $T^{1,0} \oplus T^{0,1}$. Then the *Levi form* on M is the hermitian form on $T^{1,0}$ given by $\sqrt{-1}d\theta(L, \bar{L}')$, where L and L' are in $T^{1,0}$. M is *pseudoconvex* if for some choice of θ the Levi form is non-negative.

Let $L_1, \ldots, L_{n-1}$ be a local basis for $(1, 0)$ vectorfields in a neighborhood U of $x_o \in M$ and let $\omega_1, \ldots, \omega_{n-1}$ be the dual basis of $(1, 0)$-forms. If $u \in C^\infty(U)$ we have

$$\bar{\partial}_b(u) = \sum_j \bar{L}_j(u)\bar{\omega}_j.$$

If φ is a $(0, 1)$-form on U we have

$$\bar{\partial}_b\varphi = \sum_{i<j}\Big(\bar{L}_j\varphi_i - \bar{L}_i\varphi_j + \sum_k a_{ij}^k\varphi_k\Big)\bar{\omega}_i \wedge \bar{\omega}_j$$

and

$$\bar{\partial}_b^*\varphi = -\sum_i (L_i(\varphi_i) + a_i\varphi_i),$$

where a_{ij}^k, $a_i \in C^\infty(U)$.

In general the operator $\bar{\partial}_b : \mathcal{A}_b^{0,q} \to \mathcal{A}_b^{0,q+1}$ is expressed as follows. Let $\varphi \in \mathcal{A}_b^{0,q}$ be the form given locally by:

$$\varphi = \sum_I \varphi_I\bar{\omega}_I,$$

where I is the q-tuple of integers $I = (i_1, \dots, i_q)$ with $0 < i_1 < \cdots < i_q \le n-1$ and where $\bar{\omega}_I = \bar{\omega}_{i_1} \wedge \cdots \wedge \bar{\omega}_{i_q}$. Then

$$\bar{\partial}_b \varphi = \sum_{i,I} (\bar{L}_j \varphi_I) \bar{\omega}_j \wedge \bar{\omega}_I = \sum_I \sum_{j \notin I} (\bar{L}_j \varphi_I) \bar{\omega}_j \wedge \bar{\omega}_I,$$

and

$$\bar{\partial}_b^* \varphi = -\sum_{I,H} \sum_{i \in I} \big(\epsilon_I^{iH} L_i \varphi_I + a_{(i,I)} \varphi_I\big) \bar{\omega}_H,$$

where H denotes the strictly increasing $(q-1)$-tuple and ϵ_I^{iH} is the sign of the permutation $\{i, H\} \to I$ when $\{i\} \cup H = I$ and 0 otherwise, $a_{(i,I)} \in C^\infty(U)$. The operator $\Box_b$ is defined by $\Box_b = \bar{\partial}_b \bar{\partial}_b^* + \bar{\partial}_b^* \bar{\partial}_b$.

We choose real coordinates $\{x_1, \dots, x_{2n-1}\}$ with origin at x_o such that $\theta\big(\frac{\partial}{\partial x_{2n-1}}\big) > 0$, and such that, setting $z_i = x_i + \sqrt{-1} x_{i+n-1}$ for $i = 1, \dots, n-1$, we have $L_i|_{x_o} = \frac{\partial}{\partial z_i}|_{x_o}$. Set $T = -\sqrt{-1}\frac{\partial}{\partial x_{2n-1}}$. Then the Levi form can be written as $c_{ij} = d\theta(L_i, \bar{L}_j)$ and we have

$$[L_i, \bar{L}_j] = c_{ij} T + \sum_k (a_{ij}^k L_k + b_{ij}^k \bar{L}_k),$$

where $a_{ij}^k,\ b_{ij}^k \in C^\infty(U)$.

Let $\{\xi_1, \dots, \xi_{2n-1}\}$ be the dual coordinates to $\{x_1, \dots, x_{2n-1}\}$ and $|\xi|^2 = \sum_i \xi_i^2$. Let ψ^+ and $\tilde{\psi}^+$ be a non-negative functions in $C^\infty(\{\xi \in \mathbb{R}^{2n-1} \mid |\xi| = 1\})$, with range in $[0, 1]$, such that

$$\operatorname{supp}(\psi^+) \subset \big\{|\xi| = 1 \mid \xi_{2n-1} \ge \tfrac{1}{2}|\xi'|\big\},$$
$$\operatorname{supp}(\tilde{\psi}^+) \subset \big\{|\xi| = 1 \mid \xi_{2n-1} \ge \tfrac{1}{4}|\xi'|\big\},$$

$\psi^+(\xi) = 1$ when $\xi_{2n-1} \ge \frac{3}{4}|\xi'|$ and $\tilde{\psi}^+(\xi) = 1$ when $\xi_{2n-1} \ge \frac{1}{3}|\xi'|$. Here $\xi' = (\xi_1, \dots \xi_{2n-2})$. We extend these functions to $\mathbb{R}^{2n-1}$ so that $\psi^+(\xi) = \psi^+\big(\frac{\xi}{|\xi|}\big)$ and $\tilde{\psi}^+(\xi) = \tilde{\psi}^+\big(\frac{\xi}{|\xi|}\big)$, for $|\xi| \ge 1$ and so that $\psi^+(\xi) = \tilde{\psi}^+(\xi) = 0$ for $|\xi| < \frac{1}{2}$ with $\tilde{\psi}^+(\xi) = 1$ on $\operatorname{supp}(\psi^+)$. Set $\psi^-(\xi) = \psi^+(-\xi)$, $\tilde{\psi}^-(\xi) = \tilde{\psi}^+(-\xi)$ and $(\psi^0)^2 = 1 - (\psi^+)^2 - (\psi^-)^2$, with $\psi^0 \ge 0$. Define $\tilde{\psi}^0$ so that $\tilde{\psi}^0 = 1$ on a neighborhood of $\operatorname{supp}(\psi^0)$ and so that

$$\operatorname{supp}(\tilde{\psi}^0) \subset \big\{|\xi| < 2\} \cup \{|\xi_{2n-1}| < \tfrac{3}{4}|\xi'|\big\}.$$

The operator Ψ is defined by

$$\widehat{\Psi u}(\xi) = \psi(\xi)\hat{u}(\xi),$$

The operators Ψ^+, Ψ^-, Ψ^0, $\tilde{\Psi}^+$, $\tilde{\Psi}^-$, and $\tilde{\Psi}^0$ are defined as above substituting ψ^+, ψ^-, ψ^0, $\tilde{\psi}^+$, $\tilde{\psi}^-$, and $\tilde{\psi}^0$ for ψ, respectively.

The microlocal decomposition $\varphi = \varphi^+ + \varphi^- + \varphi^0$, alluded to in the introduction, is now interpreted as follows

$$\zeta\varphi = \zeta\Psi^+\varphi + \zeta\Psi^-\varphi + \zeta\Psi^0\varphi.$$

for all $\zeta \in C_0^\infty(U)$. The microlocalization φ^+ is also sometimes interpreted as $\zeta\tilde{\Psi}^+\varphi$ and similarly with φ^- and φ^0.

The precise statement of Lemma 1.1 is then given in the following lemma, see **[K6]**.

Lemma 2.5. *Let M be a pseudoconvex CR manifold of dimension $2n-1$, let $x_o \in M$ and let λ be a real non-negative C^∞ function such that $\lambda(x_o) = 0$ and $\mathrm{Re}\big(L_i\bar{L}_j(\lambda)|_{x_o}\big)$ is positive-definite. Let $\sigma_s = e^{s\lambda}$ Then there exists a neighborhood U of x_o and $C > 0$ such that*

$$\begin{aligned}\|\sigma_s\zeta\Psi^+\varphi\|^2 \leq C\bigg(&\frac{1}{s}\max_{\operatorname{supp}\zeta} \sigma_{2s}(Q_b(\zeta\Psi^+\varphi, \zeta\Psi^+\varphi) + \|\zeta\Psi^+\varphi\|^2) \\ &+ s\max_{\operatorname{supp}\zeta} |L_i(\lambda)|^2\sigma_{2s}\|\zeta\Psi^+\varphi\|^2 + \|\Psi^+\varphi\|_{-1}^2\bigg) \qquad (2.1)\end{aligned}$$

where $\zeta \in C_0^\infty(U)$ and $\varphi \in \mathcal{A}_b^{0,1}$ with $\mathrm{supp}(\varphi) \subset U'$, *where $U' \supset \bar{U}$.*

As in **[K6]**, to generalize the above to $(0, q)$-forms we define

$$A_{IJ}(\lambda) = \mathrm{Re}\sum_{i,j,K} \epsilon_I^{iK}\epsilon_J^{jK} L_i\bar{L}_j(\lambda)$$

and

$$c_{IJ} = \sum_{i,j,K} \epsilon_I^{iK}\epsilon_J^{jK} c_{ij},$$

where K runs over all ordered $(q-1)$-tuples. Each of the coefficients ϵ_I^{iK} is either 0, 1, or -1 defined as follows. First, if $i \notin K$ we denote by (iK) the ordered q-tuple containing i and the elements of K. Then we define

$$\epsilon_I^{iK} = \begin{cases} 0 & \text{if } i \in K \\ r0 & \text{if } (iK) \neq I \\ \operatorname{sgn}\binom{iK}{I} & \text{if } (iK) = I \end{cases}$$

where $\operatorname{sgn}\binom{iK}{I}$ denotes the sign of the permutation $\{i, K\} \to I$. We define the "energy form" $Q_b(\varphi, \psi)$ by

$$Q_b(\varphi, \psi) = (\bar{\partial}_b\varphi, \bar{\partial}_b\psi) + (\bar{\partial}_b^*\varphi, \bar{\partial}_b^*\psi),$$

so that $(\Box_b\varphi, \psi) = Q_b(\varphi, \psi)$. The proof of the following is given in **[K6]**.

Lemma 2.6. *Let M be a pseudoconvex CR manifold of dimension $2n-1$, let $x_o \in M$ and let λ be a real non-negative C^∞ function such that $\lambda(x_o) = 0$ and $A_{IJ}(\lambda)|_{x_o}$ is positive-definite. Then, for $0 < q \leq n-1$, there exists a neighborhood U of x_o and $C > 0$ such that*

$$\|\sigma_s \zeta \Psi^+ \varphi\|^2 \leq C\Big(\frac{1}{s} \max_{\operatorname{supp}\zeta} \sigma_{2s}(Q_b(\zeta\Psi^+\varphi, \zeta\Psi^+\varphi) + \|\zeta\Psi^+\varphi\|^2) \\ + s \max_{\operatorname{supp}\zeta} |L_i(\lambda)|^2 \sigma_{2s} \|\zeta\Psi^+\varphi\|^2 + \|\Psi^+\varphi\|^2_{-1}\Big) \tag{2.2}$$

where $\zeta \in C_0^\infty(U)$ and $\varphi \in \mathcal{A}_b^{p,q}$, with $\operatorname{supp}(\varphi) \subset U'$, *where $U' \supset \bar{U}$.*

Again following **[K6]**, to microlocalize with Ψ^- we define the local conjugate-linear duality map $F^q : \mathcal{A}_b^{0,q} \to \mathcal{A}_b^{0,n-q-1}$ as follows. If $\varphi = \sum \varphi_I \bar{\omega}_I$ then

$$F^q(\varphi) = \sum \epsilon^{II'} \bar{\varphi}_I \bar{\omega}_{I'}, \tag{2.3}$$

where I' denotes the strictly increasing $(n-q-1)$-tuple consisting of all integers in $[1, n-1]$ which do not belong to I and $\epsilon^{II'} = \langle \bar{\omega}_I, \bar{\omega}_{I'} \rangle$. Then we have

$$\bar{\partial}_b F^q \varphi = F^{q-1} \bar{\partial}_b^* \varphi + \sum a_{IJ} \varphi_I \bar{\omega}_J$$

and

$$\bar{\partial}_b^* F^q \varphi = F^{q+1} \bar{\partial}_b \varphi + \sum b_{HK} \varphi_H \bar{\omega}_K.$$

Then we have

$$Q_b(F^q\varphi, F^q\varphi) = O(Q_b(\varphi, \varphi) + \|\varphi\|^2).$$

Thus replacing φ by $F^q\varphi$ in (3.7), we obtain

$$s \sum_{I,J} A_{I'J'}(\lambda)\sigma_s \bar{\varphi}_I, \sigma_s \bar{\varphi}_J) + \operatorname{Re} \sum_{I,J} (\sigma_s c_{I'J'} T \bar{\varphi}_I, \sigma_s \bar{\varphi}_J) + \sum_{i,J} \|\sigma_s \bar{L}_i \bar{\varphi}_J\|^2 \\ = O\big((\max_{\operatorname{supp}\varphi} \sigma_{2s})(Q_b(\varphi,\varphi) + \|\varphi\|^2) + (s \max_{\operatorname{supp}\varphi} |L_i(\lambda)| + 1)\|\sigma_s\varphi\| \sum \|\sigma_s \bar{L}_i \bar{\varphi}_J\| \\ + s^2(\max_{\operatorname{supp}\varphi} |L_i(\lambda)|^2 \sigma_{2s})\|\bar{\varphi}\|^2\big). \tag{2.4}$$

Now note that, since $\bar{T} = -T$, we have

$$\operatorname{Re} \sum_{I,J} (\sigma_s c_{I'J'} T \bar{\varphi}_I, \sigma_s \bar{\varphi}_J) = -\operatorname{Re} \sum_{I,J} (\sigma_s c_{J'I'} T \varphi_I, \sigma_s \varphi_J).$$

Hence, substituting $\zeta\Psi^-\varphi$ for φ, we can apply Lemma 2.6 and obtain the following result.

Lemma 2.7. *Let M be a pseudoconvex CR manifold of dimension $2n-1$, let $x_o \in M$ and let λ be a real non-negative C^∞ function such that $\lambda(x_o) = 0$ and $A_{I'J'}(\lambda)|_{x_o}$ is* *positive-definite. Then, for* $0 \le q < n-1$*, there exists a neighborhood* U *of* x_o *and* $C > 0$ *such that*

$$\|\sigma_s \zeta \Psi^- \varphi\|^2 \le C\Big(\frac{1}{s} \max_{\operatorname{supp}\zeta} \sigma_{2s}(Q_b(\zeta\Psi^-\varphi, \zeta\Psi^-\varphi) + \|\zeta\Psi^-\varphi\|^2)$$

$$+ s \max_{\operatorname{supp}\zeta} |L_i(\lambda)|^2 \sigma_{2s} \|\zeta\Psi^-\varphi\|^2 + \|\Psi^-\varphi\|^2_{-1}\Big) \qquad (2.5)$$

where $\zeta \in C_0^\infty(U)$ *and* $\varphi \in \mathcal{A}^{0,q}$*, with* $\operatorname{supp}(\varphi) \subset U'$*, where* $U' \supset \bar{U}$*.*

3. Isolated non-subelliptic points

In this section we will prove the following theorems.

Theorem 3.1. *Let* M *be a pseudoconvex compact CR manifold. Suppose that there exists an* $x_o \in M$ *and a neighborhood* U *of* x_o *such that* $\Box_b$ *is subelliptic on* (p,q)*-forms at each point of* $U - \{x_o\}$*. Then, for* $0 < q < n-1$*,* $\Box_b$ *on* (p,q)*-forms is hypoelliptic at all points of* U*.*

One of the main interest in the hypoellipticity of $\Box_b$ on (p,q)-forms is its consequences for the operator $\bar\partial_b$ on $(p, q-1)$-forms. If $\mathcal{H}^{p,q}$ denotes the L_2 null space of $\Box_b$ on (p,q)-forms and $S_b : L_2^{p,q} \to \mathcal{H}^{p,q}$ denotes the orthogonal projection of square-integrable (p,q)-forms on $\mathcal{H}^{p,q}$, then we have.

Corollary 3.2. *Under the assumptions of the above theorem, if* $0 \le q < n-1$ *and if* β *is a square-integrable* (p,q)*-form whose restriction to* U *is in* C^∞ *then the restriction of* $S_b\beta$ *to* U *is also in* C^∞*. Furthermore, if* $\bar\partial_b\varphi = \alpha$*, where* φ *and* α *are square-integrable* (p,q) *and* $(p,q+1)$*-forms, respectively, with* φ *orthogonal to* $\mathcal{H}^{p,q}$ *and the restriction of* α *to* U *in* C^∞ *then the restriction of* φ *to* U *is also in* C^∞*.*

The following theorem shows that if the $\bar\partial$-Neumann problem is subelliptic at each $x \ne x_o$ then it is hypoelliptic at x_o.

Theorem 3.3. *Let* X *be a complex manifold with* $\dim X \ge 2$ *and let* $\Omega \subset X$ *be a pseudoconvex domain with smooth boundary, denoted by* $b\Omega$*, such that* $\bar\Omega$ *is compact. Suppose that there exists an* $x_o \in b\Omega = M$ *and a neighborhood* U *of* x_o *such that whenever* $V \subset \bar V \subset U \cap M$ *is an open set in* M *with* $x_o \notin V$ *and* $\zeta \in C_0^\infty(V)$*. Then, for* $0 < q \le n-1$*, there exist constants* ε *and* C *such that*

$$\|\zeta\Psi^+\varphi\|_\varepsilon^2 \le C(Q_b(\varphi,\varphi) + \|\varphi\|^2), \qquad (3.1)$$

for all $\varphi \in \mathcal{A}_b^{p,q}$*. Then the* $\bar\partial$*-Neumann problem is hypoelliptic at all points of* $U \cap M$*.*

First we need the following lemma which controls the “elliptic” region.

Lemma 3.4. *If M is pseudoconvex, $U \subset M$, and $\zeta \in C_0^\infty(U)$ then there exists $C > 0$ such that*

$$\|\Psi^0 \zeta\varphi\|_1^2 \leq C(Q_b(\zeta\varphi, \zeta\varphi) + \|\varphi\|^2), \tag{3.2}$$

for all $\varphi \in \mathcal{A}_b^{p,q}$.

The lemma is a consequence of the identity

$$Q_b(\varphi, \varphi) = \sum \|\bar{L}_i \varphi_I\|^2 + \sum (c_{IJ} T\varphi_I, \varphi_J) + O(\|\varphi\|_1 \|\varphi\| + \|\varphi\|^2),$$

for $\varphi \in \mathcal{A}^{pq} \cap C_0^\infty(U)$, and of the inequality

$$\|\Psi^0 u\|_1^2 \leq C\Big(\sum \|\bar{L}_i u\|^2 + \|u\|^2\Big),$$

for all $u \in C_0^\infty(U)$.

Lemma 3.5. *Let M be a pseudoconvex CR manifold and $x_o \in M$. Then there exists $C > 0$ and for any $\delta > 0$ a neighborhood U of x_o, such that*

$$\|\zeta\varphi\|^2 \leq \delta Q_b(\zeta\varphi, \zeta\varphi) + C\|\varphi\|_{-1}^2, \tag{3.3}$$

for all $\varphi \in \mathcal{A}_b^{p,q}$ with $0 < q < n-1$ and all $\zeta \in C_0^\infty(U)$.

Proof. For $u \in C_0^\infty(U)$ we define $\Lambda^s u$ by $\widehat{\Lambda^s u}(\xi) = (1 + |\xi|^2)^{s/2}\hat{u}(\xi)$ and $\Lambda^s \varphi = \sum \Lambda^s \varphi_I$. Then, if the diameter of U is sufficiently small, we have

$$\begin{aligned}
\|\Psi^0 \zeta\varphi\|^2 &= \|\Lambda^{-1}\Psi^0\zeta\varphi\|_1^2 = \|\Psi^0 \Lambda^{-1}\zeta\varphi\|_1^2 + O(\|\varphi\|_{-1}^2) \\
&\leq C(Q_b(\zeta\Psi^0\Lambda^{-1}\varphi, \zeta\Psi^0\Lambda^{-1}\varphi) + \|\varphi\|_{-1}^2) \\
&\leq C(\|\Lambda^{-1}\bar\partial_b \zeta\varphi\|^2 + \|\Lambda^{-1}\bar\partial_b^* \zeta\varphi\|^2 + \|\varphi\|_{-1}^2) \\
&\leq \frac{\delta}{3} Q_b(\zeta\varphi, \zeta\varphi) + C\|\varphi\|_{-1}^2.
\end{aligned} \tag{3.4}$$

Next we set: $\sigma_s = \exp s(\sum |z_i|^2)$, $s \sim \frac{1}{\delta}$, and $\operatorname{diam}(U) \sim \delta$; then, from (2.2) we obtain, if δ is sufficiently small,

$$\|\zeta\Psi^+\varphi\|^2 \leq \frac{\delta}{3} Q_b(\zeta\varphi, \zeta\varphi) + C\|\varphi\|_{-1}^2. \tag{3.5}$$

Similarly from (2.5) we obtain

$$\|\zeta\Psi^-\varphi\|^2 \leq \frac{\delta}{3} Q_b(\zeta\varphi, \zeta\varphi) + C\|\varphi\|_{-1}^2. \tag{3.6}$$

Combining (3.4), (3.5), and (3.6), we obtain

$$\|\zeta\varphi\|^2 \leq \delta Q_b(\zeta\varphi, \zeta\varphi) + C\|\varphi\|_{-1}^2, \tag{3.7}$$

as desired. □

Proof of Theorem 3.2. We will show that if $\Box_b\varphi = \alpha$, where φ and α are square-integrable forms with $\alpha|_U \in C^\infty(U)$ then $\varphi|_U \in C^\infty(U)$. More precisely we will show that $\zeta\varphi \in H^s$ for all $\zeta \in C_0^\infty(U)$ and all $s \geq 0$. It will suffice to do this for ζ supported in a small neighborhood of x_o whose size depends on s. Let $\zeta', \zeta'' \in C_0^\infty(U)$ such that $\zeta' = 1$ on a neighborhood of supp ζ and $\zeta'' = 1$ on a neighborhood of supp ζ'. Then, substituting $\Lambda^s\zeta'\varphi$ for φ in (3.3) we get

$$\|\zeta\Lambda^s\zeta'\varphi\|^2 \leq \delta Q_b(\zeta\Lambda^s\zeta'\varphi, \zeta\Lambda^s\zeta'\varphi) + C\|\zeta'\varphi\|_{s-1}^2,$$

note that

$$\|\zeta\varphi\|_s^2 \leq \|\zeta\Lambda^s\zeta'\varphi\|^2 + C\|\zeta'\varphi\|_{s-1}^2.$$

Further we have

$$\begin{aligned} Q_b(\zeta\Lambda^s\zeta'\varphi, \zeta\Lambda^s\zeta'\varphi) &\leq |Q_b(\varphi, \zeta'\Lambda^s\zeta^2\Lambda^s\zeta'\varphi)| + C\|\zeta\varphi\|_s^2 + C^*\|\zeta_0\varphi\|_s^2 \\ &\leq |(\alpha, \zeta'\Lambda^s\zeta^2\Lambda^s\zeta'\varphi)| + C\|\zeta\varphi\|_s^2 + C^*\|\zeta_0\varphi\|_s^2 \\ &\leq \|\zeta\alpha\|_s^2 + C\|\zeta\varphi\|_s^2 + C^*\|\zeta_0\varphi\|_s^2, \end{aligned}$$

where $C^* = \max|d\zeta|$ and $\zeta_0 \in C_0^\infty(U)$ with $\zeta_0 = 1$ in a neighborhood of supp $d\zeta$, $\zeta' = 1$ in a neighborhood of supp ζ_0, and $\zeta_0 = 0$ in a neighborhood of x_o. Since subellipticity holds in a neighborhood of supp ζ_0 there exist $\varepsilon > 0$ such that

$$\|\zeta_0\varphi\|_s^2 \leq C(\|\zeta'\alpha\|_{s-\varepsilon}^2 + \|\varphi\|^2).$$

Hence, choosing δ sufficiently small, we obtain

$$\|\zeta\varphi\|_s^2 \leq C(\|\zeta'\alpha\|_s^2 + \|\varphi\|^2). \tag{3.8}$$

The above inequality is an apriori estimate, that is, it holds under the assumption that $\zeta\varphi \in C_0^\infty(U)$. We can now apply elliptic regularization, as developed in **[KN]**, to show that $\zeta\varphi \in H^s$ thus completing the proof. □

Proof of Theorem 3.3. The proof of this theorem is entirely analogous to the proof of Theorem 8.3 of **[K6]**. □

References

[C] D. Catlin, Subelliptic estimates for the $\bar\partial$-Neumann problem on pseudoconvex domains, Ann. of Math. 126 (1987), 131–191; Necessary conditions for the subellipticity of the $\bar\partial$-Neumann problem, Ann. of Math. 117 (1983), 147–171.

[Ch 1] M. Christ, Superlogarithmic gain implies hypoellipticity, preprint 1997.

[Ch 2] M. Christ, Spiraling and nonhypoellipticity, in: Complex Analysis and Geometry, Ohio State Univ. Math. Res. Inst. Publ. 9. Walter de Gruyter, Berlin–New York 2001, 85–101.

[Ch 3] Hypoellipticity in the infinitely degenerate regime, in: Complex Analysis and Geometry, Ohio State Univ. Math. Res. Inst. Publ. 9. Walter de Gruyter, Berlin–New York 2001, 59–84.

[D] J. P. D'Angelo, Real hypersurfaces, orders of contact and applications, Ann. of Math. 115 (1982), 615–637.

[DK] J. P. D'Angelo and J. J. Kohn, Subelliptic estimates and finite type, in: Several Complex Variables (Schneider, M. et al., eds.), Math. Sci. Res. Inst. Publ. 37, Cambridge Univ. Press, 1999, 199–232.

[F] V. S. Fediĭ, On a criterion for hypoellipticity, Math. USSR-Sb. 14 (1971), 15–45.

[K1] J. J. Kohn, Boundaries of complex manifolds, in: Proc. Conf. on Complex Analysis, Springer-Verlag, 1965, 81–94.

[K2] J. J. Kohn, Subellipticity of the $\bar{\partial}$-Neumann problem on pseudoconvex domains: sufficient conditions, Acta Math. 142 (1979), 79–122.

[K3] J. J. Kohn, Microlocalization of CR structures, in: Proc. Several Complex Variables, Hangzhou Conference 1981 (J. J. Kohn et al., eds.), Birkhäuser, Boston 1984, 29–36.

[K4] J. J. Kohn, The range of the tangential Cauchy–Riemann operator, Duke Math. J. 53 (1986), 525–545.

[K5] J. J. Kohn, Hypoellipticity of some degenerate subelliptic operators, J. Funct. Anal. 159 (1998), 203–216.

[K6] J. J. Kohn, Superlogarithmic estimates on pseudoconvex domains and CR manifolds, preprint.

[K7] J. J. Kohn, Hypoellipticity at points of finite type, in: Analysis, Geometry, Number theory: the Mathematics of Leon Ehrenpreis (Grinberg, E. L. et al., ed.), Contemp. Math. 251, Amer. Math. Soc., Providence, RI, 2000, 393–398.

[KN] J. J. Kohn and L. Nirenberg, Non-coercive boundary value problems, Comm. Pure Appl. Math. 18 (1965), 459–510.

[M] Y. Morimoto, Hypoellipticity for infinitely degenerate operators, Osaka J. Math. 24 (1963), 331–351.

[R] H. Rossi, Attaching analytic spaces to a space along a pseudo-convex boundary, in: Proc. Conf. Complex Analysis, Springer-Verlag, 1965, 242–256.

Department of Mathematics
Princeton University
Princeton, NJ 08544, U.S.A.
kohn@math.princeton.edu

Very ampleness part of Fujita's conjecture and multiplier ideal sheaves of Kohn and Nadel

*Yum-Tong Siu**

Let X be a compact complex manifold of complex dimension n with canonical line bundle K_X and L be a positive holomorphic line bundle over X. Fujita [F87] made a two-part conjecture. One part concerns effective freeness which says that for $m \geq n+1$ the line bundle $mL + K_X$ is generated by its global holomorphic sections. Another part concerns effective very ampleness which says that for $m \geq n+2$ the line bundle $mL + K_X$ is very ample. Very ampleness of $mL + K_X$ means that a basis $s_0, \ldots, s_N$ of $\Gamma(X, mL + K_X)$ over $\mathbb{C}$ defines a holomorphic embedding of X into the complex projective space $\mathbb{P}_N$ of complex dimension N.

There have been quite a number of results on the freeness part of Fujita's conjecture, including its full verification for $n \leq 4$ ([R88, Kol93, EL93, F93, Ka97]) and its partial verification for general n which requires m to be greater than a constant of order n^2 ([AS94, H97, H99]). For the very ampleness part, except for $n = 2$ in **[R88]**, the results in the literature so far give only the very ampleness of $mL + 2K_X$ for $m \geq m_n$ with some m_n depending on n and is of the order of a^n for some positive constant a ([D93, D96, ELN96, S95, S96b]). These very ampleness results use methods which cannot be applied to remove the coefficient 2 in $mL + 2K_X$.

This article contains four sections. The first section outlines a method of proving the very ampleness of $mL + K_X$ for $m \geq m_n$ with m_n depending only on n. It can actually give the global generation of q-jets for $m \geq m_{n,q}$ with $m_{n,q}$ depending only on n and q. To illustrate the method, we give only the details in the case $n = 2$. The basic idea of the method is essentially the same for both the case $n = 2$ and the case of general $n > 2$. However, in the case of general $n > 2$ the formulation used in a detailed proof is much more involved. We will work out and present the details of the case of general $n > 2$ elsewhere.

The other three sections (§2, §3, §4) consist of discussions, of a conjectural nature, on the relation of the very ampleness part of Fujita's conjecture to L^2 estimates involving derivatives and on Kohn's [Koh79] and Nadel's [N89] multiplier ideal sheaves. These discussions present ideas and problems, rather than results, in the topics under discussion.

The second section relates the very ampleness part of Fujita's conjecture to a regularity problem of $\bar{\partial}$ for locally Stein domains with L^2 estimates involving first-order derivatives.

*Partially supported by a grant from the National Science Foundation.

The third section discusses the orders of the "spikes" of a domain associated to a singular metric of a holomorphic line bundle and the application to the question of the rationality of Lelong numbers [L68, S74] which arises from the problem of the finite generation of the canonical ring of a compact manifold of general type.

The fourth section discusses the relation between Kohn's multiplier ideal sheaf [Koh79] and Nadel's multiplier ideal sheaf [N89].

1. Techniques for the very ampleness part of Fujita's conjecture

The usual method of using multiplier ideal sheaves to obtain Fujita's conjecture type results is to attempt to construct, for points P_j of X and nonnegative integer q_j ($1 \leq j \leq k$), a singular metric h for the line bundle mL over X so that

(i) the curvature current of h dominates some strictly positive smooth $(1,1)$-form on X in the sense of currents, and

(ii) the multiplier ideal sheaf $\mathcal{I}$ of the metric h (in the sense of Nadel [N89]) has the property that its zero-set is isolated at each P_j and it is contained in $\mathbf{m}_{P_j}^{q_j+1}$ $(1 \leq j \leq k)$.

Here for a point P of X, $\mathbf{m}_P$ means the maximum ideal of X at P. The multiplier ideal sheaf $\mathcal{I}$ of h (in the sense of Nadel [N89]) means the sheaf of all holomorphic function germs f on X with $h\,|f|^2$ locally integrable on X. For convenience of description we refer to a singular metric satisfying Property (i) as an *admissible singular metric*.

Once we have an admissible singular metric h, Property (i) implies the vanishing of $H^1\left(X, \mathcal{I}\left(mL+K_X\right)\right)$ by Nadel's vanishing theorem [N89]. (In the special case when the singular metric is algebraic geometrically defined, Nadel's vanishing theorem is the same as the theorem of Kawamata–Viehweg [Ka82, V82].) From the exactness of

$$\Gamma\left(X, mL+K_X\right) \rightarrow \oplus_{j=1}^{k}\left(\mathcal{O}_X / \mathbf{m}_{P_j}^{q_j+1}\right) \rightarrow H^1\left(X, \mathcal{I}\left(mL+K_X\right)\right)$$

it follows that any prescribed q_j-jets of mL at P_j ($1 \leq j \leq k$) can be simultaneously achieved by some global holomorphic section of mL over X.

The usual way of constructing the singular metric h is to use a finite number of multivalued holomorphic sections $s_1, \ldots, s_N$ of mL over X so that

$$h=\frac{1}{\sum_{j=1}^{N}\left|s_j\right|^2}.$$

Here a multivalued holomorphic section s of aL over X means that, for some positive integer p, its p-th power s^p is a holomorphic section of paL over X. We say that the multivalued holomorphic section s vanishes to order γ at a point P of X when

s^p vanishes to order $p\gamma$ at P. This definition of a multivalued holomorphic section of aL makes sense even when a is just a positive rational number. To construct the multivalued holomorphic sections $s_1, \ldots, s_N$ of mL over X, the usual way is to apply the theorem of Riemann–Roch [Hi66] either directly to X or to a subvariety of X and then form products whose factors are multivalued holomorphic sections on X and multivalued holomorphic sections extended from subvarieties of X which satisfy certain vanishing order requirements.

The reason why by this method the very ampleness part of Fujita's conjecture is so much harder to handle than its freeness part can be illustrated by the following simple calculus lemma.

(1.1) Lemma. *Let a, b, c be positive numbers. Then*

$$\int_{0<x<1, 0<y<1} \frac{dxdy}{x^a(x^b + y^c)}$$

is infinite if and only if $\max\left(a, a + b\left(1 - \frac{1}{c}\right)\right) \geq 1$.

Proof. If $a \geq 1$, then clearly

$$\int_{0<x<1, 0<y<1} \frac{dxdy}{x^a(x^b + y^c)} \geq C \int_{0<x<1} \frac{dx}{x^a} = +\infty,$$

where C is a positive constant. Now assume that $a < 1$. We distinguish among the following three cases.

Case 1. $c < 1$. We have

$$\int_{0<x<1, 0<y<1} \frac{dxdy}{x^a(x^b + y^c)} \leq \left(\int_{0<x<1} \frac{dx}{x^a}\right)\left(\int_{0<y<1} \frac{dy}{y^c}\right) < +\infty.$$

In this case,

$$a + b\left(1 - \frac{1}{c}\right) \leq a < 1$$

and

$$\max\left(a, a + b\left(1 - \frac{1}{c}\right)\right) < 1.$$

Case 2. $c = 1$. We have

$$\int_{0<x<1, 0<y<1} \frac{dxdy}{x^a(x^b + y)} = \int_{0<x<1} \left[\frac{\log(x^b + y)}{x^a}\right]_{y=0}^{y=1} dx$$

$$= \int_{0<x<1} \frac{\log\left(1 + \frac{1}{x^b}\right)}{x^a} dx < +\infty.$$

In this case,

$$\max\left(a, a+b\left(1-\frac{1}{c}\right)\right)=a<1.$$

Case 3. $c>1$. We have

$$\begin{aligned}
&\int_{0<x<1,0<y<1}\frac{dxdy}{x^a(x^b+y^c)}=+\infty\\
\Leftrightarrow\quad&\int_{0<x<1,0<y<1}\frac{dxdy}{x^a\left(x^{\frac{b}{c}}+y\right)^c}=+\infty\\
\Leftrightarrow\quad&\frac{-1}{c-1}\int_{0<x<1}\left[\frac{1}{x^a\left(x^{\frac{b}{c}}+y\right)^{c-1}}\right]_{y=0}^{y=1}dx=+\infty\\
\Leftrightarrow\quad&\frac{-1}{c-1}\int_{0<x<1}\frac{dx}{x^a\left(x^{\frac{b}{c}}+1\right)^{c-1}}+\frac{1}{c-1}\int_{0<x<1}\frac{dx}{x^a\left(x^{\frac{b}{c}}\right)^{c-1}}=+\infty\\
\Leftrightarrow\quad&\int_{0<x<1}\frac{dx}{x^a\left(x^{b\left(1-\frac{1}{c}\right)}\right)}=+\infty\\
\Leftrightarrow\quad&a+b\left(1-\frac{1}{c}\right)\geq 1.
\end{aligned}$$

□

(1.2) Corollary. *Let p and q be nonnegative integers. Let a, b, c be positive numbers. Then*

$$\int_{0<x<1,0<y<1}\frac{x^py^qdxdy}{x^a(x^b+y^c)}$$

is infinite if and only if

$$\max\left(\frac{a}{p+1},\frac{a}{p+1}+\frac{b}{p+1}\left(1-\frac{q+1}{c}\right)\right)\geq 1.$$

Proof. We use the substitution $u=x^{p+1}$ and $v=y^{q+1}$ and need only replace a by $\frac{a}{p+1}$, b by $\frac{b}{p+1}$, and c by $\frac{c}{q+1}$ in Lemma (1.1). □

(1.3) Lemma. *Let a, b, and c be positive numbers such that a and $c\left(1-\frac{\lceil a\rceil-a}{b}\right)$ are non-integers and $\lceil a\rceil-a<b<1$. Let $p_0=\lceil a-1\rceil$ and $q_0=\left\lfloor c\left(1-\frac{\lceil a\rceil-a}{b}\right)\right\rfloor$. Then on $\mathbb{C}^2$ with coordinates z, w, the multiplier ideal sheaf for the singular metric*

$$\frac{1}{|z|^{2a}\left(|z|^{2b}+|w|^{2c}\right)}$$

is generated by z^{p_0+1} *and* $z^{p_0}w^{q_0}$. *Here* $\lfloor\cdot\rfloor$ *denotes the round-down (i.e., the largest integer not exceeding that number) and* $\lceil\cdot\rceil$ *denotes the round-up (i.e., the smallest integer not less than that number).*

Proof. Let $x = |z|^2$ and $y = |w|^2$. Let

$$d\Lambda = \frac{1}{|z|^{2a}\left(|z|^{2b} + |w|^{2c}\right)}\left(\frac{\sqrt{-1}}{\pi}dz \wedge d\bar{z}\right) \wedge \left(\frac{\sqrt{-1}}{\pi}dw \wedge d\bar{w}\right).$$

Let p and q be nonnegative integers. The integration of

$$\left|z^p w^q\right|^2 d\Lambda \tag{1.3.1}$$

over the circle $|z| = 1$ and $|w| = 1$ yields

$$\frac{x^p y^q dxdy}{x^a\left(x^b + y^c\right)}$$

so that by Lemma (1.2) the local integrability of (1.3.1) at 0 is equivalent to

$$\begin{cases} a < p+1 \\ \frac{a}{p+1} + \frac{b}{p+1}\left(1 - \frac{q+1}{c}\right) < 1. \end{cases}$$

In other words, the local integrability of (1.3.1) at 0 is equivalent to

$$\begin{cases} p > a-1 \\ q > c\left(1 - \frac{p+1-a}{b}\right) - 1. \end{cases}$$

We now distinguish among the following three cases.

Case 1. $p \geq p_0 + 1 = \lceil a-1 \rceil + 1$. Since a is not an integer, in this case we have $\lceil a-1 \rceil + 1 > a - 1 + 1 = a$ and, in particular, $p > a - 1$. Moreover, in this case

$$\begin{aligned} c\left(1 - \frac{p+1-a}{b}\right) - 1 &\leq c\left(1 - \frac{\lceil a-1 \rceil - (a-1) + 1}{b}\right) - 1 \\ &< c\left(1 - \frac{1}{b}\right) - 1 < -1, \end{aligned}$$

because $1 - \frac{1}{b} < 0$ from $b < 1$ and $c > 0$. Hence the condition

$$q > c\left(1 - \frac{p+1-a}{b}\right) - 1$$

is always satisfied for any nonnegative integer q. Thus (1.3.1) is always locally integrable at 0 when $p \geq p_0 + 1$.

Case 2. $p = p_0 = \lceil a - 1 \rceil$. In this case we also have $p > a - 1$, because a is not an integer. Since

$$c\left(1 - \frac{p+1-a}{b}\right) = c\left(1 - \frac{\lceil a \rceil - a}{b}\right),$$

for any nonnegative integer q, the condition

$$q > c\left(1 - \frac{p+1-a}{b}\right) - 1$$

is equivalent to

$$q \geq \left\lfloor c\left(1 - \frac{\lceil a \rceil - a}{b}\right)\right\rfloor,$$

because $c\left(1 - \frac{\lceil a \rceil - a}{b}\right)$ is not an integer. Thus (1.3.1) is locally integrable at 0 when $p = p_0$ and $q \geq q_0$.

Case 3. $p < p_0$. Then $p + 1 \leq \lceil a - 1 \rceil < a$, because a is not an integer. Thus (1.3.1) is not locally integrable at 0 when $p < p_0$.

Combining the three cases, we conclude that (1.3.1) is locally integrable at 0 if either $p > p_0$ or

$$\begin{cases} p = p_0 \\ q \geq q_0 \end{cases}$$

Let $f(z, w)$ be a holomorphic function germ on $\mathbb{C}^2$ at $(z, w) = (0, 0)$. We can write

$$f = g_1(z, w) + z^{p_0} g_2(w) + z^{p_0} w^{q_0} h_1(z, w) + z^{p_0+1} h_2(z, w),$$

where

$$g_1(z, w) = \sum_{p=0}^{p_0-1} \sum_{q=0}^{\infty} a_{p,q} z^p w^q,$$

$$g_2(w) = \sum_{q=0}^{q_0-1} a_{p_0,q} w^q,$$

with $a_{p,q} \in \mathbb{C}$, and $h_1(z, w)$ and $h_2(z, w)$ are holomorphic function germs on $\mathbb{C}^2$ at $(z, w) = (0, 0)$. To finish the proof of the lemma, it suffices to show that $|f(z, w)|^2 \, d\Lambda$ is locally integrable at 0 if and only if both $g_1(z, w)$ and $g_2(w)$ are identically zero. Since $\left|z^{p_0+1}\right|^2 d\Lambda$ and $|z^{p_0} w^{q_0}|^2 \, d\Lambda$ are both locally integrable at 0, without loss of generality we can assume that both $h_1(z, w)$ and $h_2(z, w)$ are identically zero. Assume that one of $g_1(z, w)$ and $g_2(w)$ is not identically zero. We are going to show that $|f(z, w)|^2 \, d\Lambda$ is not locally integrable at 0. We distinguish between the following two cases.

Case (i). $g_1(z, w)$ is not identically zero. Suppose $|f|^2 \, d\Lambda$ is integrable on some open neighborhood U of 0 in $\mathbb{C}^2$ and we are going to derive a contradiction. Let p_1 be the smallest integer $0 \leq p < p_0$ such that $\sum_{q=0}^{\infty} a_{p,q} w^q$ is not identically zero. Then there exists an open neighborhood W of some point of $U \cap \{z = 0, \ w \neq 0\}$ in U such that

$$|f(z, w)| \geq A \, |z|^{p_1}$$

on W, where A is some positive number. Since $p_1 < p_0 = \lceil a - 1 \rceil$, it follows that $p_1 < a - 1$, because the inequality $p_1 \geq a - 1$ would have implied $p_1 \geq \lceil a - 1 \rceil$ from the fact that p_1 is an integer. Hence

$$\frac{|z^{p_1}|^2}{|z|^{2a}} \left(\frac{\sqrt{-1}}{\pi} dz \wedge d\bar{z} \right) \wedge \left(\frac{\sqrt{-1}}{\pi} dw \wedge d\bar{w} \right)$$

is not integrable on W and, in particular, $|z^{p_1}|^2 \, d\Lambda$ is not integrable on W. It follows that $|f|^2 \, d\Lambda$ is not integrable on W, which contradicts $W \subset U$.

Case (ii). $g_1(z, w)$ is identically zero. Then $g_2(w)$ is not identically zero. Let q_2 be the smallest integer $0 \leq q < q_0$ such that $a_{p_0,q}$ is nonzero. Then

$$|f(z, w)| \geq B \, |z|^{p_0} \, |w|^{q_2}$$

on some open neighborhood of 0 in $\mathbb{C}^2$, where B is some positive number. Since $|z^{p_0} w^{q_2}|^2 \, d\Lambda$ is not locally integrable at 0 from $q_2 < p_0$, it follows that $|f|^2 \, d\Lambda$ is not locally integrable at 0. □

In Lemma (1.3) the multiplier ideal sheaf of the singular metric

$$\frac{1}{|z|^{2a} \left(|z|^{2b} + |w|^{2c} \right)}$$

can also be described as being generated by monomials $z^p w^q$ so that (p, q) is no less than (p_0, q_0) in the lexicographical ordering. For convenience of description we refer to such a kind of ideal sheaf as a *lexicographical ideal sheaf.* We also use the corresponding description *lexicographical ideal* for an ideal instead of an ideal sheaf.

To continue our explanation why by such a method the very ampleness part is harder than the freeness part, for the sake of simplicity we confine ourselves to the case of $n = 2$. Fix a point P_0 in X. We first use the theorem of Riemann–Roch to get a multivalued holomorphic section s of some multiple of L over X which vanishes to a certain order at P_0. Let C be the 1-dimensional component of the zero-set of the multiplier ideal sheaf of $\frac{1}{|s|^2}$. We need only consider the case when C contains P_0. We apply the theorem of Riemann–Roch to get a multivalued section $\tilde{t}$ of some multiple of $L|C$ vanishing to a certain order at P_0 on C. Then we extend $\tilde{t}$ to a multivalued holomorphic section of the same multiple of L over all of X. The vanishing order of t at P_0 on X in general may be a very small positive number even though the vanishing order of $\tilde{t}$ at P_0 on C is very large. For example, for a local coordinate system (z, w) of X centered at P_0, the curve C can be locally given by $z = 0$ and t may be locally given by $\left(w^{Np} + z\right)^{\frac{1}{N}}$ so that the vanishing order of $t|C$ is p at P_0 on C but the vanishing order of t at p on X is only $\frac{1}{N}$. The multiplier ideal sheaf of the singular metric $\frac{1}{|st|^2}$ resembles the multiplier ideal sheaf of the singular metric

$$\frac{1}{|z|^{2a}\left(|z|^{2b} + |w|^{2c}\right)}$$

with $b > 0$ very small. So we end up with a multiplier ideal sheaf which behaves like a lexicographical ideal sheaf at P_0. To make the support of a lexicographical ideal sheaf isolated at a point, it is possible if we require only that the ideal sheaf is contained in $\mathbf{m}_{P_0}$, but it is impossible in general if we require that the ideal sheaf is contained in $\mathbf{m}^q_{P_0}$ for some $q \geq 2$. For this reason this method can give results on the freeness but not the very ampleness part.

To handle the very ampleness part of Fujita's conjecture, in the method of this paper we consider not just the multiplier ideal sheaf $\mathcal{I}$ of an admissible singular metric, but also other coherent ideal sheaves constructed as the middle term of a short exact sequence of multiplier ideal sheaves of singular metrics. By applying Nadel's vanishing theorem [N89] to multiplier ideal sheaves of admissible singular metrics, from the short exact sequences we get also the corresponding vanishing of cohomology for the coherent ideal sheaves which are the middle terms of short exact sequences. These coherent ideal sheaves can be constructed in such a way that their zero-sets are isolated at the prescribed points and they are contained in the prescribed powers of the maximum ideals at those prescribed points. The construction is essentially modelled on recovering a high power of the maximum ideal of a regular local ring from the lexicographical ideals.

For the case of complex dimension two, the method yields the following theorem.

(1.4) Theorem. *Let X be a compact complex surface and L be a positive holomorphic line bundle over X. Let $P_1, \ldots, P_\kappa$ be a finite number of distinct points in X and $q_1, \ldots, q_\kappa$ be positive integers. Let $m \geq 1 + \left(2 + \sum_{j=1}^{\kappa} q_j\right)^2$. Then, for any prescribed q_j-jet at P_j for $1 \leq j \leq \kappa$, there exists an element $\Gamma(X, mL + K_X)$ whose q_j-jet at P_j is the prescribed one for $1 \leq j \leq \kappa$.*

We will carry out the proof only for the case $\kappa = 1$, because the modification needed to get the case of general κ is completely straightforward and is obvious from the proof for $p = 1$.

For $n = 2$ the construction of the desired coherent ideal sheaf needs only one step. The following trivial lemma highlights why the construction of the desired coherent ideal sheaf is so much simpler. In the case of general n the zero dimension condition has to be replaced by the condition of complex codimension at least two and additional conditions on the quotient sheaf will be needed.

(1.5) Lemma. *Let $\mathcal{I} \subset \mathcal{J}$ be coherent ideal sheaves on a compact complex surface X such that the support of the quotient $\mathcal{J}/\mathcal{I}$ has dimension zero. Let L be a holomorphic line bundle over X. If $H^p(X, \mathcal{I}(L+K_X)) = 0$ for $p \geq 1$, then $H^p(X, \mathcal{J}(L+K_X)) = 0$ for $p \geq 1$.*

Proof. From the exact sequence

$$0 \to \mathcal{I}(L+K_X) \to \mathcal{J}(L+K_X) \to (\mathcal{J}/\mathcal{I})(L+K_X) \to 0$$

one has the exact sequence

$$H^p(X, \mathcal{I}(L+K_X)) \to H^p\left(X, \mathcal{J}(L+K_X)\right) \to H^p\left(X, (\mathcal{J}/\mathcal{I})(L+K_X)\right).$$

Since the support of the quotient $\mathcal{J}/\mathcal{I}$ has dimension zero, it follows that

$$H^p\left(X, (\mathcal{J}/\mathcal{I})(L+K_X)\right) = 0$$

for $p \geq 1$. The conclusion of the lemma follows. □

(1.6) *Proof of Theorem* (1.4) *for* $p = 1$. To prove Theorem (1.4) for the case $p = 1$, we denote P_1 by P_0 and $q_1 + 1$ by q. Choose arbitrarily a rational number $0 < \eta < \frac{1}{2}$ which will later be required to be very small and whose smallness we will specify toward the end of this proof. Let s be a multivalued holomorphic section of L over X such that the vanishing order ord_{s,P_0} of s at P_0 is at least $1 - \eta$. Let

$$\hat{q} = \frac{q+1}{\mathrm{ord}_{s,P_0}}.$$

Let $h_\alpha = \frac{1}{|s|^{2\alpha}}$. Let $\mathcal{I}_\alpha$ be the multiplier ideal sheaf of the metric h_α of the $\mathbb{Q}$-bundle αL for $0 \leq \alpha \leq \hat{q}$. There exist

(i) a finite subset F of X and

(ii) a finite number of complex curves $C_1, \ldots, C_\ell$ in X such that

  (a) on $X - F$ the ideal sheaf $\mathcal{I}_\alpha$ is the ideal sheaf of a divisor of the form $\sum_{j=1}^{\ell} a_{\alpha,j} C_j$ for $0 \leq \alpha \leq \hat{q}$, where $a_{\alpha,j}$ $(1 \leq j \leq \ell, 0 \leq \alpha \leq \hat{q})$ are nonnegative integers,

  (b) $a_{\alpha,j} \leq a_{\beta,j}$ for $0 \leq \alpha < \beta \leq \hat{q}$ and $1 \leq j \leq \ell$, and

  (c) $a_{\hat{q},j} > 0$ for $1 \leq j \leq \ell$.

There exists a sequence

$$0 = \alpha_0 < \alpha_1 < \alpha_2 < \cdots < \alpha_k \leq \hat{q}$$

such that

$$\sum_{j=1}^{\ell} a_{\alpha,j} = \begin{cases} 0 & \text{for } 0 < \alpha < \alpha_1 \\ \nu & \text{for } \alpha_\nu \leq \alpha < \alpha_{\nu+1} \text{ and } 1 \leq \nu < k \\ k & \text{for } \alpha_k \leq \alpha \leq \hat{q}. \end{cases}$$

As a current

$$\frac{\sqrt{-1}}{2\pi}\partial\bar{\partial}\log|s|^{2\hat{q}} \geq \sum_{j=1}^{\ell} a_{\hat{q},j}\left[C_j\right],$$

where $\left[C_j\right]$ is the closed positive (1, 1)-current on X defined by integration over the regular part of C_j. Since the Lelong number of the current

$$\frac{\sqrt{-1}}{2\pi}\partial\bar{\partial}\log|s|^{2\hat{q}}$$

at P_0 is $q+1$, it follows that

$$\sum_{j=1}^{\ell} a_{\hat{q},j}\left(\text{mult}_{P_0} C_j\right) \leq q+1$$

and, in particular,

$$\sum_{j=1}^{\ell} \text{mult}_{P_0} C_j \leq q+1,$$

where $\text{mult}_{P_0} C_j$ is the multiplicity of C_j at P_0.

Let Δ be the open disk in $\mathbb{C}$ of radius 1 centered at the origin. Take an open neighborhood U of P_0 in X and a local holomorphic family of nonsingular complex lines E_λ in U parametrized by $\lambda \in \Delta$ such that

(i) E_λ contains P_0 when $\lambda = 0$,

(ii) $E_\lambda \cap \left(\bigcup_{j=1}^{\ell} C_j\right)$ is 0-dimensional for $\lambda \in \Delta$,

(iii) E_λ intersects $\bigcup_{j=1}^{\ell} C_j$ only at the regular points of $\bigcup_{j=1}^{\ell} C_j$ and all the intersections are normal crossing for $\lambda \in \Delta - 0$, and

(iv) the number of points in $E_\lambda \cap \left(\bigcup_{j=1}^{\ell} C_j\right)$ is no more than $q+1$ for $\lambda \in \Delta - 0$.

For $\lambda \in \Delta$ by using the theorem of Riemann–Roch we can find a multivalued holomorphic section $\tilde{t}_\lambda$ of $(q+1)L|\left(\bigcup_{j=1}^{\ell} C_j\right)$ over $\left(\bigcup_{j=1}^{\ell} C_j\right)$ so that the sum of the vanishing order of t_λ at all the points of $E_\lambda \cap \left(\bigcup_{j=1}^{\ell} C_j\right)$ is at least $1-\eta$ for $\lambda \in \Delta - 0$ and the dependence of t_λ on the parameter λ is holomorphic for $\lambda \in \Delta$. We can extend $\tilde{t}_\lambda$ to a multivalued holomorphic section t_λ of $(q+1)L$ over X so that the dependence of t_λ on the parameter λ is holomorphic for $\lambda \in \Delta$. Let $t = t_0$.

Let $\sigma_1, \ldots, \sigma_{N_0}$ be holomorphic sections of $m_0 L$ over X without common zeros in X for some large positive integer m_0. Let $\sigma = \sum_{j=1}^{N_0} |\sigma_j|^{\frac{2}{m_0}}$. We fix some positive rational number ϵ, δ. Consider the metric

$$h_{\alpha,\beta} = \frac{1}{|s|^{2(\alpha-\epsilon)}\sigma^{\epsilon}\left(|s|^{2\delta}\sigma^{\beta(q+1)-\delta} + |t|^{2\beta}\right)}$$

of the $\mathbb{Q}$-bundle $(\alpha + \beta(q+1))\,L$ over X. The range of α being considered for $h_{\alpha,\beta}$ is $0 \leq \alpha \leq \hat{q}$. Let $\mathcal{J}_\alpha$ be the multiplier ideal sheaf of $h_{\alpha,\beta}$ when $\beta = \frac{q}{1-2\eta}$. Let $\mathcal{I}'_{\alpha_\nu}$ be the ideal sheaf of $\sum_{1 \leq j \leq \ell} a_{\alpha_\nu, j} C_j$ for $0 \leq \nu \leq k$. For some sufficiently small appropriately chosen ϵ, δ we have

$$\mathcal{J}_{\alpha_\nu} \subset \mathbf{m}_{P_0}^q + \mathcal{I}'_{\alpha_\nu}$$

and

$$\mathcal{J}_{\alpha_\nu} \subset \mathcal{I}_{\alpha_{\nu-1}}$$

for $1 \leq \nu \leq k$. Let

$$\mathcal{J}'_{\alpha_\nu} = \mathcal{J}_{\alpha_\nu} + \mathcal{I}'_{\alpha_\nu}.$$

From

$$\mathcal{I}_{\alpha_\nu} \subset \mathcal{I}'_{\alpha_\nu}$$

it follows that

$$\mathcal{I}'_{\alpha_\nu} \subset \mathcal{J}'_{\alpha_\nu} \subset \mathbf{m}_{P_0}^q + \mathcal{I}'_{\alpha_\nu}$$

and

$$\mathcal{J}'_{\alpha_\nu} = (\mathcal{J}'_{\alpha_\nu} \cap \mathbf{m}_{P_0}^q) + \mathcal{I}'_{\alpha_\nu}.$$

From

$$\mathcal{J}'_{\alpha_\nu} \subset \mathcal{I}_{\alpha_{\nu-1}} \subset \mathcal{I}'_{\alpha_{\nu-1}}$$

it follows that

$$\mathcal{J}'_{\alpha_\nu} \subset \left(\mathcal{I}'_{\alpha_{\nu-1}} \cap \mathbf{m}_{P_0}^q\right) + \mathcal{I}'_{\alpha_\nu}.$$

Fix an integer $m > \hat{q} + \frac{q}{1-2\eta}(q+1)$. Since the dimension of the support of the quotient

$$\left(\left(\mathcal{I}'_{\alpha_{\nu-1}} \cap \mathbf{m}_{P_0}^q\right) + \mathcal{I}'_{\alpha_\nu}\right) / \mathcal{J}_{\alpha_\nu}$$

is zero, it follows from

$$H^p\left(X, \mathcal{J}_{\alpha_\nu}(mL + K_X)\right) = 0$$

for $p \geq 1$ that

(1.6.1) $$H^p\left(X, \left(\left(\mathcal{I}'_{\alpha_{\nu-1}} \cap \mathbf{m}_{P_0}^q\right) + \mathcal{I}'_{\alpha_\nu}\right)(mL + K_X)\right) = 0$$

for $p \geq 1$ and $1 \leq \nu \leq k$.

We have the isomorphism

$$(\mathcal{I}'_{\alpha_{\nu-1}} \cap \mathbf{m}^q_{P_0})/(\mathcal{I}'_{\alpha_\nu} \cap \mathbf{m}^q_{P_0}) \to ((\mathcal{I}'_{\alpha_{\nu-1}} \cap \mathbf{m}^q_{P_0}) + \mathcal{I}'_{\alpha_\nu})/\mathcal{I}'_{\alpha_\nu}, \tag{1.6.2}$$

because

$$(\mathcal{I}'_{\alpha_{\nu-1}} \cap \mathbf{m}^q_{P_0}) \cap \mathcal{I}'_{\alpha_\nu} = \mathcal{I}'_{\alpha_\nu} \cap \mathbf{m}^q_{P_0}.$$

Since

$$\mathcal{I}_{\alpha_\nu} \subset \mathcal{I}'_{\alpha_\nu}$$

and the support of

$$\mathcal{I}'_{\alpha_\nu}/\mathcal{I}_{\alpha_\nu}$$

has dimension zero, it follows from

$$H^p(X, \mathcal{I}_{\alpha_\nu}(mL + K_X)) = 0$$

for $p \geq 1$ and $0 \leq \nu \leq k$ that

$$H^p(X, \mathcal{I}'_{\alpha_\nu}(mL + K_X)) = 0 \tag{1.6.3}$$

for $p \geq 1$ and $0 \leq \nu \leq k$. By (1.6.1) and (1.6.3), we have

$$H^p(X, (((\mathcal{I}'_{\alpha_{\nu-1}} \cap \mathbf{m}^q_{P_0}) + \mathcal{I}'_{\alpha_\nu})/\mathcal{I}'_{\alpha_\nu})(mL + K_X)) = 0$$

for $p \geq 1$ and $1 \leq \nu \leq k$. From the isomorphism (1.6.2) it follows that

$$H^p(X, ((\mathcal{I}'_{\alpha_{\nu-1}} \cap \mathbf{m}^q_{P_0})/(\mathcal{I}'_{\alpha_\nu} \cap \mathbf{m}^q_{P_0}))(mL + K_X)) = 0 \tag{1.6.4}$$

for $p \geq 1$ and $1 \leq \nu \leq k$. We have

$$\mathcal{I}_{\alpha_k} \cap \mathbf{m}^q_{P_0} = \mathcal{I}_{\alpha_k},$$

because

$$\mathcal{I}_{\alpha_k} \subset \mathbf{m}^q_{P_0}.$$

Hence

$$H^p(X, (\mathcal{I}_{\alpha_k} \cap \mathbf{m}^q_{P_0})(mL + K_X)) = 0$$

for $p \geq 1$ and $0 \leq \nu \leq k$. Since

$$\mathcal{I}_{\alpha_k} \cap \mathbf{m}^q_{P_0} \subset \mathcal{I}'_{\alpha_k} \cap \mathbf{m}^q_{P_0}$$

and the support of

$$(\mathcal{I}'_{\alpha_k} \cap \mathbf{m}^q_{P_0})/(\mathcal{I}_{\alpha_k} \cap \mathbf{m}^q_{P_0})$$

is of dimension zero, it follows that

$$H^p(X, (\mathcal{I}'_{\alpha_k} \cap \mathbf{m}^q_{P_0})(mL + K_X)) = 0 \tag{1.6.5}$$

for $p \geq 1$ and $0 \leq \nu \leq k$. By using the exact sequence

$$0 \to \mathcal{I}'_{\alpha_\nu} \cap \mathbf{m}^q_{P_0} \to \mathcal{I}'_{\alpha_{\nu-1}} \cap \mathbf{m}^q_{P_0} \to \left(\mathcal{I}'_{\alpha_{\nu-1}} \cap \mathbf{m}^q_{P_0}\right) / \left(\mathcal{I}'_{\alpha_\nu} \cap \mathbf{m}^q_{P_0}\right) \to 0$$

and descending induction on $1 \leq \nu \leq k$, we conclude from (1.6.4) and (1.6.5) that

$$H^p\left(X, \left(\mathcal{I}'_{\alpha_\nu} \cap \mathbf{m}^q_{P_0}\right)(mL + K_X)\right) = 0 \tag{1.6.6}$$

for $p \geq 1$ and $0 \leq \nu \leq k$. Since $\mathcal{I}'_{\alpha_0} = \mathcal{O}_X$, the case $\nu = 0$ of (1.6.6) yields

$$H^p\left(X, \mathbf{m}^q_{P_0}(mL + K_X)\right) = 0 \tag{1.6.7}$$

for $p \geq 1$. Now choose the positive number η to be so small that

$$\hat{q} + \frac{q}{1-2\eta}(q+1) \leq \frac{1}{1-2\eta}(q+1)^2 < 1 + (q+1)^2.$$

From the exact sequence

$$0 \to \mathbf{m}^q_{P_0} \to \mathcal{O}_X \to \mathcal{O}_X / \mathbf{m}^q_{P_0} \to 0$$

and (1.6.7) it follows that $\Gamma(X, mL + K_X)$ generate all the $(q-1)$-jets at P_0 for $m \geq 1 + (q+1)^2$. This finishes the proof of Theorem (1.4) for the case $\kappa = 1$. □

Remark. In the proof of Theorem (1.4) there is no attempt to give a sharp lower bound for m. Even without introducing any new techniques, the proof in (1.6) could be easily modified to improve somewhat the lower bound for m given in (1.4).

2. Relation between very ampleness and L^2 estimates involving derivatives

The very ampleness part of Fujita's conjecture is related, in the following way, to a regularity problem of $\bar\partial$ for locally Stein domains with L^2 estimates involving derivatives. Now we no longer restrict ourselves to the surface case and allow n to be any integer ≥ 2.

Let P_0 and Q_0 be two distinct points of X. For the freeness part of Fujita's conjecture one produces an admissible singular metric $h = e^{-\varphi}$ of $m_0 L$ whose multiplier ideal sheaf $\mathcal{I}$ has an isolated zero at P_0 and Q_0 and is contained in $\mathbf{m}_{P_0}$ and $\mathbf{m}_{Q_0}$. Then we use Nadel's vanishing theorem to get the vanishing of $H^1(X, \mathcal{I}(mL + K_X))$ for $m \geq m_0$. It means that for any $\bar\partial$-closed mL-valued $(n,1)$-form f on X with

$$\int_X |f|^2 e^{-\varphi} < \infty,$$

we can solve the $\bar{\partial}$-equation $\bar{\partial} u = f$ on X for an mL-valued $(n, 0)$-form u on X with L^2 estimates

$$\int_X |u|^2 e^{-\varphi} < \infty$$

with weight $e^{-\varphi}$. Here $|f|^2$ is the pointwise L^2-norm square with respect to any smooth metric of X.

If we can solve the $\bar{\partial}$-equation $\bar{\partial} u = f$ on X with L^2 estimates with weight $e^{-\varphi}$ for derivatives up to order 1, then we can conclude the very ampleness of $mL + K_X$. More precisely, if for any $\bar{\partial}$-closed mL-valued $(n, 1)$-form f on X with

$$\int_X \left(|f|^2 + |\nabla f|^2\right) e^{-\varphi} < \infty, \tag{2.1}$$

we can solve the $\bar{\partial}$-equation $\bar{\partial} u = f$ on X for an mL-valued $(n, 0)$-form u on X such that

$$\int_X \left(|u|^2 + |\nabla u|^2\right) e^{-\varphi} < \infty, \tag{2.2}$$

then we can conclude the very ampleness of $mL + K_X$, where ∇ is the covariant derivative in the direction of $(1, 0)$ with respect to any smooth metric of X. (Note that the finiteness condition of the above two integrals is independent of the choice of the smooth metric of X.) The reason is as follows.

Take any open neighborhood U of P_0 in X and any $g \in \Gamma(U, mL + K_X)$. Let $0 \leq \rho \leq 1$ be a smooth function supported on U which is identically 1 on some open neighborhood U' of P_0 in U. Let $f = \bar{\partial}(\rho g)$. Clearly f satisfies (2.1). If we have a solution u of $\bar{\partial} u = f$ which satisfies (2.2), then u is holomorphic on U' and belongs to $\mathbf{m}_{P_0}^2$. Hence $g - u$ is an element of $\Gamma(X, mL + K_X)$ whose 1-jet at P_0 agrees with the 1-jet of g at P_0. Thus we can conclude the very ampleness of $mL + K_X$.

Let $e^{-\psi}$ be a smooth metric of L with strictly positive curvature on X. There is a technique introduced in [Koh73] to get L^2 estimates involving derivatives on pseudoconvex domains in $\mathbb{C}^n$. It uses an additional weight function to take care of the error terms arising from the commutation of $\bar{\partial}$ and ∇ and the commutation of $\bar{\partial}^*$ and ∇. The reason why in the above discussion we increase m_0 to m is that we hope to use $e^{-(m-m_0)\psi}$ to take the place of the additional weight function in the technique of [Koh73].

One difficulty in applying the technique of [Koh73] to our problem at hand is that the process of taking the commutation of $\bar{\partial}$ and ∇ (and also of $\bar{\partial}^*$ and ∇) introduces the differentiation of φ. We have no way of estimating the derivatives of φ. One way of avoiding the difficulty of differentiating φ is as follows.

We consider the total bundle space of the dual L^* of L and introduce the open subset Ω consisting of all elements of L^* whose length with respect to the metric h^{-1} is < 1. Let $\pi : L^* \to X$ be the bundle projection and let w be its fiber coordinate.

Let $\tilde{f} = (\pi^* f) \wedge \frac{dw}{w}$. Then $\tilde{f}$ is a scalar-valued $(n+1, 1)$-form on L^*. Since f is an $(n, 1)$-form with value in mL, we know that $\tilde{f}$ vanishes to order $m-1$ at the zero-section Z of the line bundle $\pi : L^* \to X$. Instead of the $\bar{\partial}$-equation $\bar{\partial} u = f$ on X, we use the $\bar{\partial}$-equation $\bar{\partial}\tilde{u} = \tilde{f}$ on Ω to solve for the $(n+1, 0)$-form $\tilde{u}$ on Ω with

$$\int_\Omega |u|^2 < \infty.$$

Since $\tilde{f}$ and $\tilde{u}$ are scalar-valued forms instead of line-bundle-valued forms, we no longer have to use any weight function from the metric of the line bundle. The role of the original weight function $e^{-\varphi}$ is now replaced by the volume of the fiber disk inside Ω. The intersection of Ω with a fiber is the disk $|w|^2 < e^{-\varphi}$ and

$$\frac{1}{2p\pi} \int_{|w|^2 < e^{-\varphi}} \left| w^{p-1} \right|^2 \sqrt{-1} dw \wedge d\bar{w} = e^{-\varphi} \tag{2.3}$$

for any positive integer p.

To be able to solve the $\bar{\partial}$-equation $\bar{\partial}\tilde{u} = \tilde{f}$ on Ω with L^2 estimates involving derivatives, the technique of [Koh73] requires an additional weight and we need a strictly plurisubharmonic function on Ω or at least on $\Omega - Z$. To get such a function we introduce $\tau = |w|^2 e^{\psi}$, where $e^{-\psi}$ is the smooth metric of L over X with strictly positive curvature on X. We use τ to help us get the extra weight function to try to solve the $\bar{\partial}$-equation $\bar{\partial}\tilde{u} = \tilde{f}$ on Ω with L^2 estimates for derivatives up to order 1 (where the derivative is computed with respect to the local coordinate system of L^* consisting of w and the pullback of a local coordinate system of X).

One hopes that for the estimate involving derivatives of order up to 1 some appropriate weight function constructed from τ can take care of the commutation terms by using the techniques of [Koh73]. The function τ vanishes at the zero-section Z of the line bundle $\pi : L^* \to X$ and is strictly plurisubharmonic only outside Z. The vanishing of $\tilde{f}$ to order m at Z will be used to take care of the vanishing order and the lack of plurisubharmonicity of τ at Z. The main difficulty is how to make sure that m can be chosen to depend only on n.

Once one gets a solution $\tilde{u}$, one takes its power series expansion with respect to the fiber coordinate w and singles out the term $uw^{m-1}dw$. Then u is an mL-valued $(n, 0)$-form on X which solves the $\bar{\partial}$-equation $\bar{\partial} u = f$ on X. The L^2 estimate of u with weight $e^{-m\varphi}$ involving derivatives up to order 1 comes from (2.3). The singularity of $e^{-\varphi}$ is now reflected in the non-relative-compactness of Ω in L^*. The domain Ω has infinite "spikes" at the points of X where φ becomes $-\infty$.

The idea of introducing Ω to transform line-bundle-valued forms to scalar-valued forms traces back to Grauert [G62].

3. Spikes and rationality of Lelong numbers

Consider a metric

$$h = \frac{1}{\sum_{j=1}^{N} |s_j|^2}$$

of a holomorphic line bundle L over a compact complex manifold X of complex dimension n, where s_j $(1 \leq j \leq N)$ are multivalued holomorphic sections of L over X. There are two ways in which h may fail to be a smooth metric with positive curvature. The first one is that $s_1, \ldots, s_N$ all vanish at some point of X. The second one is that at some point P of X there exists no subset of $n+1$ elements in $\{s_1, \ldots, s_N\}$ which can serve as a local homogeneous coordinate system at P. Let Ω be the domain in the dual bundle L^* of L consisting of all elements whose length is < 1 with respect to the metric h^{-1} of L^*. The first way of failure corresponds to the base point set of mL when ${s_1}^m, \ldots, {s_N}^m$ form a basis of $\Gamma(X, mL)$ over $\mathbb{C}$. It also corresponds to the "spikes" of the domain Ω in L^*. The second way of failure corresponds to the non-strictly pseudoconvex points in the boundary of Ω.

Non-strictly pseudoconvex boundary points, in the context of the regularity of the Kohn solution, have been extensively studied (for example, [B84, B92, BS90, BS91, BS93, C82, C83, C87, Ch91, Chr96, D'AK99, FKM90, Ki91, Koh73, Koh79, Koh97, Koh98, Koh00]). However, no theory has yet been developed for "spikes". We outline here a problem which could be used as a guide for the development of a theory for "spikes".

One long outstanding conjecture in algebraic geometry is the finite generation of the canonical ring $R(X, K_X)$ of a compact complex projective algebraic manifold X of general type. General type means that the dimension of $\Gamma(X, mK_X)$ dominates $c\,m^n$ when m is sufficiently large, where c is a positive constant (independent of m) and n is the complex dimension of X. For a holomorphic line bundle L over X the ring $R(X, L)$ is defined by

$$R(X, L) = \bigcup_{m=1}^{\infty} \Gamma(X, mL).$$

Let

$$s_1^{(m)}, \ldots, s_{q_m}^{(m)} \in \Gamma(X, mL)$$

be a basis over $\mathbb{C}$ and let $\epsilon_m > 0$ $(1 \leq m < \infty)$ be a sequence of positive numbers chosen to be so rapidly decreasing that

$$\Phi_L := \sum_{m=1}^{\infty} \epsilon_m \sum_{j=1}^{q_m} \left| s_j^{(m)} \right|^{\frac{2}{m}}$$

is convergent on X as a metric of L^*. If the ring $R(X, L)$ is finitely generated, the vanishing orders of Φ_L must be rational numbers, or more precisely, the Lelong numbers of the closed positive (1, 1)-current

$$\frac{\sqrt{-1}}{2\pi}\partial\bar{\partial}\log\Phi_L$$

must be rational numbers (see [L68, S74] for the theory of Lelong numbers and closed positive currents).

One possible approach to obtaining the rationality of Lelong numbers for the case $L = K_X$ is to use a property of Kohn's multiplier ideal sheaf involving certain differential operators. The setting of Kohn's multiplier ideal sheaf [Koh79] is a bounded domain D in $\mathbb{C}^n$ with smooth weakly pseudoconvex boundary defined by $r < 0$ so that dr is nowhere zero on the boundary ∂D of D. The stalk, at a point P, of Kohn's multiplier ideal sheaf $\mathcal{I}^{(1)}$ for (0, 1)-forms consists of all smooth function germs F at P such that there exist an open neighborhood U of P in $\mathbb{C}^n$ and positive numbers ϵ and C (both depending on F) satisfying

$$|||Fg|||_\epsilon^2 \leq C\left(\|\bar{\partial}g\|^2 + \|\bar{\partial}^*g\|^2 + \|g\|^2\right)$$

for every (0, 1)-form g supported on $U \cap \bar{D}$ which is in the domain of $\bar{\partial}$ and $\bar{\partial}^*$. Here $|||\cdot|||_\epsilon$ is the L^2 norm on D involving derivatives up to order ϵ in the boundary tangential direction of D, $\|\cdot\|$ is the usual L^2 norm on D without involving any derivatives, and $\bar{\partial}^*$ is the actual adjoint of $\bar{\partial}$ with respect to $\|\cdot\|$. Likewise, the stalk, at a point P, of Kohn's multiplier module sheaf $\mathcal{M}^{(1)}$ for (0, 1)-forms consists of all smooth germs of (1, 0)-forms ω at P such that there exist an open neighborhood U of P in $\mathbb{C}^n$ and positive numbers ϵ and C (both depending on ω) satisfying

$$|||g\cdot\omega|||_\epsilon^2 \leq C\left(\|\bar{\partial}g\|^2 + \|\bar{\partial}^*g\|^2 + \|g\|^2\right)$$

for every (0, 1)-form g supported on $U \cap \bar{D}$ which is in the domain of $\bar{\partial}$ and $\bar{\partial}^*$. Here $g\cdot\omega$ is the pointwise contraction of the (0, 1)-form g with the (1, 0)-form ω. There are a number of properties of $\mathcal{I}^{(1)}$ and $\mathcal{M}^{(1)}$. For example, r is in $\mathcal{I}^{(1)}$; the contraction of $\partial\bar{\partial}r$ with any smooth (1, 0)-form orthogonal to ∂r is in $\mathcal{M}^{(1)}$; and any smooth function germ whose absolute value is dominated by the absolute value of some positive power of an element of $\mathcal{I}^{(1)}$ is itself in $\mathcal{I}^{(1)}$.

There are two important properties of $\mathcal{I}^{(1)}$ and $\mathcal{M}^{(1)}$ which are most relevant to our discussion. The first one is that the ∂ of an element of $\mathcal{I}^{(1)}$ is in $\mathcal{M}^{(1)}$. The second one is that, if $\omega_1, \ldots, \omega_n$ belong to $\mathcal{M}^{(1)}$, then the coefficient of the $(n, 0)$-form $\omega_1 \wedge \cdots \wedge \omega_n$ belongs to $\mathcal{I}^{(1)}$. These two properties together give us the property that some nonlinear differentiation process involving a number of elements of $\mathcal{I}^{(1)}$ again yields an element of $\mathcal{I}^{(1)}$.

One hopes that this particular property of $\mathcal{I}^{(1)}$ involving a nonlinear differentiation process would be able to give us a system of linear equations with coefficients in $\mathbb{Q}$ whose unknowns are the vanishing orders of $\mathcal{I}^{(1)}$. When the system of linear equations is nondegenerate enough, one could conclude the rationality of such vanishing orders.

Kohn's multiplier ideal sheaves are for non-strictly pseudoconvex boundary points. The analogous argument for their counterpart for "spikes" might yield the rationality of the Lelong numbers of

$$\frac{\sqrt{-1}}{2\pi}\partial\bar{\partial}\log\Phi_L$$

(or the vanishing orders of Φ_L) when $L = K_X$. One guide for the development of a theory for the "spikes" analogous to the theory for the non-strictly pseudoconvex points is to prove the rationality of the Lelong numbers of

$$\frac{\sqrt{-1}}{2\pi}\partial\bar{\partial}\log\Phi_L$$

when $L = K_X$.

This argument could work only for the case $L = K_X$, because of the process of using $\omega_1 \wedge \cdots \wedge \omega_n$ for $\omega_1, \ldots, \omega_n \in \mathcal{M}^{(1)}$. This corresponds to the fact that if we take $(1,0)$ covariant differentiation of smooth sections $t_1, \ldots, t_n$ of mK_X over X to get sections $Dt_1, \ldots, Dt_n$ of $K_X \otimes T_X^*$ over X, the exterior product $Dt_1 \wedge \cdots \wedge Dt_n$ is a section of $(nm+1)K_X$. Here T_X^* is the cotangent bundle of X.

So far as the behavior at the base point set of X is concerned, $\log \Phi_L$ for $L = K_X$ may be linked in some way to the potential in the Monge–Ampére equation for the "Kähler–Einstein" current for X.

4. Relation between Kohn's and Nadel's multipliers

Nadel's multiplier ideal sheaf is defined so that the $\bar{\partial}$-equation $\bar{\partial}u = f$ could be solved when f belongs to Nadel's multiplier ideal sheaf. On the other hand, Kohn's multiplier ideal sheaf consists of multipliers so that, after a "test function" for the estimate needed for the solution of the $\bar{\partial}$-equation is multiplied by the multiplier, the product satisfies the estimate. Nadel's multiplier ideal sheaf is for the right-hand side of the $\bar{\partial}$-equation and Kohn's multiplier ideal sheaf is for the "test function" of the $\bar{\partial}$-equation. It is natural to expect some form of duality between the two kinds of multiplier ideal sheaves.

Consider the weakly pseudoconvex domain D in $\mathbb{C}^{n+1}$ with coordinates $(w, z) = (w, z_1, \ldots, z_n)$ whose boundary is given by

$$\operatorname{Re} w = \sum_{j=1}^{k} \left|f_j(z)\right|^2,$$

where $f_j(z)$ are holomorphic functions in z $(1 \leq j \leq k)$. Suppose we apply to $\mathbb{C}^n$ (with coordinates z) a monoidal transformation with a regular center and get $\pi : Y \to \mathbb{C}^n$. For example, we can blow up a single point in $\mathbb{C}^n$. We consider the weakly pseudoconvex domain $\tilde{D}$ in $\mathbb{C} \times Y$ whose boundary is defined by

$$\operatorname{Re} w = \sum_{j=1}^{k} \left|f_j \circ \pi\right|^2.$$

When the boundary of D does not contain any positive-dimensional complex-analytic subvarieties, the boundary of $\tilde{D}$ may contain positive-dimensional complex-analytic subvarieties. One hopes that, by considering the pullback to $\tilde{D}$ of (0, 1)-forms from D, in the situation of complex-analytic subvarieties in the boundary one can arrive at a general formulation of the condition for the right-hand side for the $\bar{\partial}$-equation for regularity to hold. Such a formulation would mix the concept of Kohn's multiplier with the concept of Nadel's multiplier and might yield a better understanding of the relation between the two.

One possible application of this kind of consideration is to apply monoidal transformations with nonsingular centers to reduce the set of holomorphic functions $f_1, \ldots, f_k$ to a more standard situation where the Hölder estimates for $\bar{\partial}$-equation can be more easily investigated. At this point $\bar{\partial}$ Hölder estimates for weakly pseudoconvex domains are known only for very special cases such as a simultaneously diagonalizable Levi form ([FKM90]).

References

[AS94] U. Angehrn and Y.-T. Siu, Effective freeness and separation of points for adjoint bundles, Invent. Math. 122 (1995), 291–308.

[B84] D. Barrett, Irregularity of the Bergman projection on a smooth bounded domain in $\mathbf{C}^2$, Ann. of Math. 119 (1984), 431–436.

[B92] D. Barrett, Behavior of the Bergman projection on the Diederich-Fornaess worm, Acta Math. 168 (1992), 1–10.

[BS90] H. Boas and E. Straube, Equivalence of regularity for the Bergman projection and the $\bar{\partial}$-Neumann operator, Manuscripta Math. 67 (1990), 25–66.

[BS91] H. Boas and E. Straube, Sobolev estimates for the $\bar{\partial}$-Neumann operator in domains in $\mathbf{C}^n$ admitting a defining function that is plurisubharmonic on the boundary, Math. Z. 206 (1991), 81–88.

[BS93] H. Boas and E. Straube, De Rham cohomology of manifolds containing the points of infinite type, and Sobolev estimates for the $\bar{\partial}$-Neumann problem, J. Geom. Analysis 3 (1993), 225–235.

[C82] D. Catlin, Global regularity for the $\bar{\partial}$-Neumann problem, Proc. Sympos. Pure Math. 41 (1982), 39–49.

[C83] D. Catlin, Necessary conditions for the subellipticity of the $\bar{\partial}$-Neumann problem, Ann. of Math. 117 (1983), 147–171.

[7C87] D. Catlin, Subelliptic estimates for the $\bar{\partial}$-Neumann problem on pseudoconvex domains, Ann. of Math. 126 (1987), 131–192.

[Ch91] S.-C. Chen, Global regularity of the $\bar{\partial}$-Neumann problem in dimension two, Proc. Sympos. Pure Math. 52 (1991), 55–61.

[Chr96] M. Christ, Global C^∞ irregularity of the $\bar{\partial}$-Neumann problem for worm domains, J. Amer. Math. Soc. 9 (1996), 1171–1185.

[D'AK99] J. D'Angelo and J. J. Kohn, Subelliptic estimates and finite type, in: Several complex variables (Berkeley, CA, 1995–1996), Math. Sci. Res. Inst. Publ. 37, Cambridge Univ. Press, Cambridge 1999, 199–232.

[D93] J.-P. Demailly, A numerical criterion for very ampleness, J. Differential Geom. 37 (1993), 323–374.

[D96] J.-P. Demailly, Effective bounds for very ample line bundles, Invent. Math. 124 (1996), 243–261.

[EL93] L. Ein and R. Lazarsfeld, Global generation of pluricanonical and adjoint linear series on smooth projective threefolds, J. Amer. Math. Soc. 6 (1993), 875–903.

[ELN96] Lawrence Ein, Robert Lazarsfeld, and Michael Nakamaye, Zero-estimates, intersection theory, and a theorem of Demailly, in: Higher-dimensional complex varieties (M. Andreatta and Th. Peternell, eds.), Walter de Gruyter, Berlin, New York 1996, 183–207.

[FKM90] C. Fefferman, J. J. Kohn, and M. Machedon, Hölder estimates on CR manifolds with a diagonalizable Levi form, Adv. Math. 84 (1990), 1–90.

[F87] T. Fujita, On polarized manifolds whose adjoint bundles are not semipositive, in: Algebraic Geometry, Sendai, Adv. Stud. Pure Math. 10 (1987), 167–178.

[F93] T. Fujita, Remarks on Ein-Lazarsfeld criterion of spannedness of adjoint bundles of polarized threefolds, preprint 1993.

[G62] H. Grauert, Über Modifikationen und exzeptionelle analytische Mengen, Math. Ann. 146 (1962), 331–368.

[H97] S. Helmke, On Fujita's conjecture, Duke Math. J. 88 (1997), 201–216.

[H99] S. Helmke, On global generation of adjoint linear systems, Math. Ann. 313 (1999), 635–652.

[Hi66] F. Hirzebruch, Topological Methods in Algebraic Geometry, Springer-Verlag, Berlin, Heidelberg, New York 1966.

[Ka82] Y. Kawamata, A generalization of the Kodaira-Ramanujam's vanishing theorem, Math. Ann. 261 (1982), 43-46.

[Ka97] Y. Kawamata, On Fujita's freeness conjecture for 3-folds and 4-folds, Math. Ann. 308 (1997), 491–505.

[Ki91] C. Kiselman, A study of the Bergman projection in certain Hartogs domains, Proc. Sympos. Pure Math. 52 (1991), 219–231.

[Koh73] J. J. Kohn, Global regularity for $\bar\partial$ on weakly pseudo-convex manifolds, Trans. Amer. Math. Soc. 181 (1973), 273–292.

[Koh79] J. J. Kohn, Subellipticity of the $\bar\partial$-Neumann problem on pseudo-convex domains: sufficient conditions, Acta Math. 142 (1979), 79–122.

[Koh97] J. J. Kohn, Quantitative estimates for global regularity, in: Analysis and geometry in several complex variables (Katata, 1997), 97–128, Trends Math., Birkhäuser, Boston, MA, 1999.

[Koh98] J. J. Kohn, Hypoellipticity of some degenerate subelliptic operators, J. Funct. Anal. 159 (1998), 203–216.

[Koh00] J. J. Kohn, Hypoellipticity at points of infinite type, in: Analysis, Geometry, Number Theory: The Mathematics of Leon Ehrenpreis (E. I. Grinberg et al., eds.), Contemp. Math. 251, Amer. Math. Soc., Providence, RI, 2000, 393–398.

[Kol93] J. Kollár, Effective base point freeness, Math. Ann. 296 (1993), 595–605.

[L68] P. Lelong, Fonctions plurisousharmoniques et formes différentielles positives, Gordon & Breach, Paris, London, New York (distributed by Dunod éditeur, Paris) 1968.

[N89] A. Nadel, Multiplier ideal sheaves and the existence of Kähler-Einstein metrics of positive scalar curvature, Proc. Natl. Acad. Sci. USA 86, 7299–7300 (1989), and Ann. of Math. 132 (1989), 549–596.

[R88] I. Reider, Vector bundles of rank 2 and linear systems on algebraic surfaces, Ann. of Math. 127, 309–316 (1988).

[Si74] Y.-T. Siu, Analyticity of sets associated to Lelong numbers and the extension of closed positive currents, Invent. Math. 27 (1974), 53–156.

[S95] Y.-T. Siu, Very ampleness criterion of double adjoints of ample line bundles, in: Modern methods in complex analysis (Princeton, NJ, 1992), Ann. of Math. Stud. 137, Princeton Univ. Press, Princeton, NJ, 1995, 291–318.

[S96a] Y.-T. Siu, The Fujita conjecture and the extension theorem of Ohsawa-Takegoshi, in: Geometric Complex Analysis (Junjiro Noguchi et al., eds.), World Scientific, Singapore, New Jersey, London, Hong Kong 1996, 577–592.

[S96b] Y.-T. Siu, Effective very ampleness, Invent. Math. 124 (1996), 563–571.

[V82] E. Viehweg, Vanishing theorems, J. Reine Angew. Math. 335 (1982), 1–8.

Department of Mathematics, Harvard University, Cambridge, MA 02138, U.S.A.

E-mail: `siu@math.harvard.edu`

Bei Fragen zur Produktsicherheit wenden Sie sich bitte an:
If you have any questions regarding product safety,
please contact:

Walter de Gruyter GmbH
Genthiner Straße 13
10785 Berlin
productsafety@degruyterbrill.com